Ernst Schering Research Foundation Workshop 13
Assessment of the Use of Single Cytochrome P450
Enzymes in Drug Research

Ernst Schering Research Foundation Workshop

Editors: Günter Stock
Ursula-F. Habenicht

Vol. 7
Basic Mechanisms Controlling Term and Preterm Birth
Editors: K. Chwalisz, R. E. Garfield

Vol. 8
Health Care 2010
Editors: C. Bezold, K. Knabner

Vol. 9
Sex Steroids and Bone
Editors: R. Ziegler, J. Pfeilschifter, M. Bräutigam

Vol. 10
Nongenotoxic Carcinogenesis
Editors: A. Cockburn, L. Smith

Vol. 11
Cell Culture in Pharmceutical Research
Editors: N. E. Fusening, H. Graf

Supplement 1
Molecular and Cellular Endocrinology of the Testis
Editors: G. Verhoeven, U.-F. Habenicht

Vol. 12
Interactions Between Adjuvants, Agrochemicals and Target Organisms
Editors: P. J. Holloway, R. Rees, D. Stock

Vol. 13
Assessment of the Use of Single Cytochrome P450 Enzymes
in Drug Research
Editors: M. R. Waterman, M. Hildebrand

Ernst Schering Research Foundation
Workshop 13

Assessment of the Use of Single Cytochrome P450 Enzymes in Drug Research

M. R. Waterman, M. Hildebrand
Editors

With 44 Figures

Springer-Verlag Berlin Heidelberg GmbH

ISBN 978-3-662-03021-9   ISBN 978-3-662-03019-6 (eBook)
DOI 10.1007/978-3-662-03019-6

© Springer-Verlag Berlin Heidelberg 1994
Originally published by Springer-Verlag Berlin Heidelberg New York in 1994
Softcover reprint of the hardcover 1st edition   1994

Typesetting: Data conversion by Springer-Verlag

21/3130–5 4 3 2 1 0 – Printed on acid-free paper

Preface

Pharmacokinetics has become an integral part of drug development, but can also offer a large variety of tools to optimize and rationalize early drug research.

Investigations of the metabolism of drugs is of high importance both in scientific and economic terms for pharmaceutical companies. A long known and well established example of a metabolically improved compound is the synthesis of ethinylestradiol by Inhoffen as a stabilized estrogen that has been a very successful compound on world's pharmaceutical market. Ernst Schering Research Foundation's parent has been interested in this subject due to the extensive work done especially on the field of steroids. In addition, the use of specific enzymatic properties is not only of interest for those involved in pharmacokinetics, but has been used as a very important tool to synthesize drugs with complex molecular structures and specific substitution patterns. Furthermore, there has also been a long pharmacological interest in several specific research areas, for example development of aromatase inhibition as pharmacotherapeutic approach.

Metabolic research and especially that which focuses on the monooxygenase cytochrome P450 system has gone through dramatic progressive changes within the last decade. First, availability of highly sophisticated analytical techniques has contributed to increased molecular understanding of biotransformation pathways. Nowadays designer drug are computer composable focussing on metabolic stability. The rapid development of recombinant enzymes has, in many cases, replaced the intensive work on P450 isolation from naturally occurring sources. Recombinant P450s can be expressed in cell lines and thus

Fig. 1. The participants of the workshop

made available for reproducible and constant use addressing a large number of scientific questions.

This rapid progress and the active involvement of Ernst Schering Research Foundation's parent in all possible applications of single enzymatic properties was the rationale for the organization of an Ernst Schering Research Foundation workshop on the "Assessment of the use of single cytochrome P450 enzymes in drug research" which was held in Berlin from March 23–25, 1994. This was the 13th meeting of this kind organized by the Ernst Schering Research Foundation and offered a superb forum for exchange of recent scientific research data and built a bridge between the fascinating new possibilities offered by biotechnology and its applications for conventional and new questions in drug development and research. The present book contains the proceedings of the symposium, describes the present position of such research and should serve as a stimulus for future work in both academia and industry on cytochrome P450 enzymes.

M. Waterman
M. Hildebrand

Table of Contents

1   Xenobiotic Metabolism:
    From Intact Biosystem to Single Enzymes
    M. Hildebrand, M. Hümpel, H. Gieschen, and C. Kraus . . .   1

2   Regulation of Xenobiotic-Metabolizing Cytochromes P450
    F. J. Gonzalez   21

3   Cytochrome P450 in Human Drug
    Metabolism: How Much Is Predictable?
    U. A. Meyer   43

4   Metabolic Activation of Anticancer
    Oxazaphosphorines by Cytochrome P450s:
    Development of a Model for Cancer Gene Therapy
    L. Chen and D. J. Waxman   57

5   The Use of Bacteria for Cytochrome P450 Expression
    M. R. Waterman   81

6   Expression of Cytochromes P450
    in Yeast: Practical Aspects
    D. Pompon, G. Truan, A. Bellamine, and P. Urban   97

7   Human Cells as an Expression System for Cytochromes P450
    C. L. Crespi, R. Langenbach, H. V. Gelboin, F. J. Gonzalez,
    and B. W. Penman   111

8   Metabolic Reactions and Recombinant Isoenzymes
     of Cytochrome P450: Information Generated
     and Value for Pharmaceutical Development
     R. E. Tynes 135

9   The Importance of Cytochrome P450 3A Enzymes
     in Drug Metabolism
     F. P. Guengerich, E. M. J. Gillam, M. V. Martin, T. Baba,
     B.-R. Kim, T. Shimada, K. D. Raney, and C.-H. Yun 161

10   Strategies for Verifying the Involvement
     of Specific Cytochrome P450 Enzymes
     in Xenobiotic Metabolism
     P. H. Beaune, S. Lecoeur, C. Belloc, A. Lemoine,
     D. Pompon, J.-C. Gautier, and H. Kroemer 187

11   Regulation of Cytochromes P450
     by Substrate Interactions
     M. Ingelman-Sundberg, A. Zhukov, S. Mkrtchian,
     and E. Eliasson 195

12   Strategies for the Use of Single Cytochrome P450 Enzymes
     in Drug Research and Future Prospects
     J. Doehmer 213

Subject Index 225

List of Contributors

T. Baba
Department of Biochemistry and Center in Molecular Toxicology,
Vanderbilt University School of Medicine, Nashville, TN 37232, USA

P. H. Beaune
Inserm U 75, Université René Descartes, 156 rue de Vaugirard,
75730 Paris cedex 15, France

A. Bellamine
Centre de Génétique Moléculaire, UPR 2420 du CNRS,
l'Université Pierre-et-Marie-Curie, 91198 Gif-sur-Yvette cédex, France

C. Belloc
Inserm U 75, Université René Descartes, 156 rue de Vaugirard,
75730 Paris cedex 15, France

L. Chen
Division of Cell and Molecular Biology, Department of Biology,
Boston University, Boston, MA 02215, USA

C. L. Crespi
GENTEST Corporation, 6 Henshaw Street, Woburn, MA 01801, USA

J. Doehmer
Institut für Toxikologie und Umwelthygiene, TU München,
Lazarettstr. 62, 80636 Munich, Germany

E. Eliasson
Department of Medical Biochemistry and Biophysics, Karolinska Institute,
171 77 Stockholm, Sweden

J.-C. Gautier
Inserm U 75, Université René Descartes, 156 rue de Vaugirard,
75730 Paris cedex 15, France

H. V. Gelboin
Laboratory of Molecular Carcinogenesis, National Cancer Institute,
National Institutes of Health, Bethesda, MD 20892, USA

H. Gieschen
Institute of Pharmacokinetics, Schering AG, 13342 Berlin, Germany

E. M. J. Gillam
Department of Biochemistry and Center in Molecular Toxicology,
Vanderbilt University School of Medicine, Nashville, TN 37232, USA

F. J. Gonzalez
Laboratory of Molecular Carcinogenesis, National Cancer Institute,
National Institutes of Health, Bethesda, MD 20892, USA

F. P. Guengerich
Department of Biochemistry and Center in Molecular Toxicology,
Vanderbilt University School of Medicine, Nashville, TN 37232, USA

M. Hildebrand
Institute of Pharmacokinetics, Schering AG, 13342 Berlin, Germany

M. Hümpel
Institute of Pharmacokinetics, Schering AG, 13342 Berlin, Germany

M. Ingelman-Sundberg
Department of Medical Biochemistry and Biophysics, Karolinska Institute,
171 77 Stockholm, Sweden

B.-R. Kim
Department of Biochemistry and Center in Molecular Toxicology,
Vanderbilt University School of Medicine, Nashville, TN 37232, USA

C. Kraus
Institute of Pharmacokinetics, Schering AG, 13342 Berlin, Germany

H. Kroemer
Dr. Margarete Fischer-Bosch-Institut für klinische Pharmakologie,
Auerbachstr. 110, 70376 Stuttgart, Germany

R. Langenbach
National Institute of Environmental Health Sciences, P.O. Box 12233,
Research Triangle Park, NC 27709, USA

S. Lecoeur
Inserm U 75, Université René Descartes, 156 rue de Vaugirard,
75730 Paris cedex 15, France

A. Lemoine
Inserm U 75, Université René Descartes, 156 rue de Vaugirard,
75730 Paris cedex 15, France

M. V. Martin
Department of Biochemistry and Center in Molecular Toxicology
Vanderbilt University School of Medicine, Nashville, TN 37232, USA

U. A. Meyer
Biozentrum of the University of Basel, Department of Pharmacology,
Klingelbergstr. 70, 4056 Basel, Switzerland

S. Mkrtchian
Department of Medical Biochemistry and Biophysics, Karolinska Institute,
171 77 Stockholm, Sweden

B. W. Penman
GENTEST Corporation, 6 Henshaw Street, Woburn, MA 01801, USA

D. Pompon
Centre de Génétique Moléculaire, UPR 2420 du CNRS
l'Université Pierre-et-Marie-Curie, 91198 Gif-sur-Yvette cédex, France

K. D. Raney
Department of Biochemistry and Center in Molecular Toxicology
Vanderbilt University School of Medicine, Nashville, TN 37232, USA

T. Shimada
Department of Biochemistry, and Center in Molecular Toxicology,
Vanderbilt University School of Medicine, Nashville, TN 37232, USA

G. Truan
Centre de Génétique Moléculaire, UPR 2420 du CNRS,
l'Université Pierre-et-Marie-Curie, 91198 Gif-sur-Yvette cédex, France

R. E. Tynes
Drug Metabolism/Biopharmaceutics, Drug Safety Department,
Sandoz Pharma Ltd., 4002 Basel, Switzerland

P. Urban
Centre de Génétique Moléculaire, UPR 2420 du CNRS,
l'Université Pierre-et-Marie-Curie, 91198 Gif-sur-Yvette cédex, France

M. R. Waterman
Department of Biochemistry, Vanderbilt University School of Medicine,
Nashville, TN 37232, USA

D. J. Waxman
Division of Cell and Molecular Biology, Department of Biology,
Boston University, Boston, MA 02215,USA

C.-H. Yun
Department of Biochemistry and Center in Molecular Toxicology,
Vanderbilt University School of Medicine, Nashville, TN 37232, USA

A. Zhukov
Department of Medical Biochemistry and Biophysics, Karolinska Institute,
171 77 Stockholm, Sweden

1 Xenobiotic Metabolism: From Intact Biosystem to Single Enzymes

M. Hildebrand, M. Hümpel, H. Gieschen, and C. Kraus

1.1	Metabolism	1
1.1.1	Development, Functions, and Reactions	1
1.1.2	Aims of Studies and Investigational Tools	2
1.2	Practical Use of Different Models	5
1.2.1	Metabolic Stabilization of Prostacyclin Mimetics	5
1.2.2	Metabolic Profiling of Ergoline Derivatives	7
1.2.3	V79 Cell Lines with Single P450 Enzymes as Screening Tool	9
1.3	Evaluative Comparison of Different Models	14
1.4	Discussion and Conclusion	15
References		17

1.1 Metabolism

1.1.1 Development, Functions, and Reactions

The evolutionary process has yielded an endogenous system in living organisms that is able to transform and functionalize compounds for life requirements, deactivate them in the postfunctional phase, and thus provide the equipment for handling endo- and xenobiotics which we term metabolic capacity.

The reactions of biotransformation can be attributed to two types: phase I and phase II steps. Phase I reactions are also described as functionalizing reactions and comprise direct, enzymatically mediated changes at the molecule, such as oxidations, reductions, and hydrolytical cleavages. In the case of oxidative reactions the mono-oxygenase cytochrome P450 system plays an important role.

Phase II reactions are conjugations in which the parent compound, if structurally suitable, or a phase I metabolite is paired with another small or middle-sized molecule, for example, glucuronic acid, sulfate, acetic acid, glutathion, or an amino acid such as glycine.

All phase I and II reactions are mediated by a series of more or less specific enzymes with different structure and activity requirements, which are often expressed in different subtypes. The large variability of possible reactions together with the involvement of different enzymes again verifies the complexity of functional differentiation in intact organisms and often complicates the investigation and understanding of biotransformation. Furthermore, chemical instability may lead to an additional nonenzymatic portion of degradation products.

Metabolic reactions lead in general to a detoxification of xenobiotics. This is reached primarily by an increase in hydrophilicity due to chemical reactions at the foreign compound, which gives the metabolite a different distribution behavior in the biosystem and improves its excretability. Sometimes metabolic transformations lead to an activation of compounds by producing highly reactive intermediates. Such activated products can be of interest if they are able to react with endogenous structures and disturb their functionality.

1.1.2 Aims of Studies and Investigational Methods

The main focus of metabolic studies is to complete the pharmacokinetic characterization of a new chemical entity, which summarizes the fate of the compound in the body by describing absorption, distribution, metabolism, and excretion. The main goal of metabolic investigation is certainly to elucidate the biodegradation in man as completely as possible. This task may be very difficult, especially in the case of pharmacologically highly potent drugs that are administered in low doses, in the microgram range, or with drugs that are subject to extensive biodegradation, sometimes yielding more than 15 or 20 products. Furthermore, we must consider methodological problems when working in the complete biosystem of man. Primary sources for metabolite studies are the excreta (predominantly urine and to a lesser extent the feces) and sometimes plasma. A complete picture of products of biodegradation can be obtained only by the use of radiolabeled material, and this is limited generally to a single dose study in man, from which all in vivo data must be generated. As the quantity of metabolites is often insufficient for isolation and elucidation, alternative sources for the generation of metabolites must be used. This leads either to animal or in vitro models.

A second, equally important rationale for metabolic studies is the toxicological risk assessment for new chemical entities, which is performed on the basis of animal studies in different species. In this context it is necessary to show a at least qualitative similarity of biotransformation pathways in the rodents and nonrodents used in comparison to man. This is done to verify that all possible toxicophores are present in the animals giving the complete risk picture. In the case of the singular occurrence of a toxic metabolite either in man or in an animal species used for toxicological evaluation risk assessment is critical. This situation leads to a second dilemma in metabolic research and the necessity to look for reliable, highly predictive in vitro alternatives. This is because a large number of toxicological studies are performed prior to administration in man, and the use of in vivo data for comparison of the metabolic profile can thus be fully validated only ex post facto.

Metabolic research can also contribute to a better understanding of the pharmacological activity in the case of the formation of active metabolites. Sometimes, if other pharmacokinetic, pharmaceutical, and physicochemical properties of active metabolites are interesting, the search for such derivatives may set the starting point for the development of a better pharmacotherapeutic alternative.

Another piece of the complex mosaic is elucidated when we consider the synthesis of a large number of structurally closely related compounds which exhibit a certain pharmacodynamic property in vitro or in animal models. Here metabolic stability tested in appropriate in vivo or in vitro systems can become a selection criterion for further development if either a reasonable pharmacological differentiation is not possible, or many compounds are subject to immediate biodegradation.

Finally, metabolic interactions are an important issue for safe and effective drug therapy. If two drugs which are used concomitantly compete for a single enzyme, saturation is possible, which may lead to increased systemic drug levels and unwanted effects (Cholerton et al. 1992). Also in a broader sense inherited absence or deficiency of certain enzymes (Gonzalez and Idle 1994), which is the subject of pharmaco-genetic studies, may lead to an altered metabolic pattern in specific subpopulations followed by increased or reduced plasma levels of the active compound.

The tools for metabolic studies start with in vivo experiments to be conducted in whole biosystems, such as animals or man after different routes of administration. The route of administration is preferably intra-vasal to avoid any problems of incomplete absorption; as mentioned above, the use of radiolabeled material provides the best opportunity to qualify and quantify the various products by chromatographic methods in complex biomatrices.

Ex vivo methods address primarily the central organ of biodegrada-tion, the liver. Nevertheless, it must be considered that a number of other tissues and organs sometimes exhibit very specific enzyme sys-tems competent for biotransformation (e.g., lungs in case of 15-OH-prostaglandin dehydrogenase). As regards the liver, its isolation and recirculating perfusion is an established method for small animals, such as the rat and guinea pig. The metabolic pattern can be studied in perfusate, bile (if secreted), and liver tissue (Meijer et al. 1981).

The use of liver slices (Barr et al. 1991) and homogenates disturbs the integrity of the organ but enables testing of several compounds in a single approach. This is also true for hepatocytes, which are obtained by collagenase perfusion of the liver. Various methods have been tested to cultivate these cells in a physiological matrix (e.g., together with epithe-lial cells) for a longer time of experiment (Coundouris et al. 1993) or in a permanent culture. This method is also applicable for human material (Guillouzo et al. 1985).

Microsomes are the metabolically highly active subcellular fractions of hepatocytes; they contain a large number of drug-metabolizing enzymes and can be stored.

In the past the use of single enzymes was limited to a time-consum-ing isolation procedure; however, advances in biotechnology have fa-cilitated the continuous availability of a definite enzyme expressed in heterologous systems (Crespi 1991; Doehmer and Oesch 1991; Gonza-lez et al. 1991a,b).

1.2 Practical Use of Different Models

1.2.1 Metabolic Stabilization of Prostacyclin Mimetics

Prostacyclin is the most potent inhibitor of thrombocyte aggregation and is thus therapeutically interesting. However, this endogenous compound strictly follows one principle of nature: synthesize the mediator when and where it is needed, and then get rid of it. This contradicts therapeutic use. Prostacyclin is both chemically and metabolically instable and is immediately biodegraded by a number of metabolic attacks in both side chains. The goal of our research was to obtain more stable analogues that retain the pharmacological potency. Therefore structural modifications were made, yielding iloprost, cicaprost, and eptaloprost as a series of metabolically increasingly stable compounds (Fig. 1). Iloprost is subject mainly to β-oxidative degradation of the upper side chain, resulting in the tetranor metabolites, which have been found in all animals species in vivo, in perfusion models, and in hepatocytes (Hildebrand 1992). The dinor intermediates are not seen. Some minor metabolic activity concerns the lower side chain in terms of hydroxylation.

Based upon this experience cicaprost was synthesized with an oxygen atom at C-4 to block β-oxidation. The compound was pharmacologically active in cardiovascular indications at doses of 5–15 µg in man and exhibited high metabolic stability (Stürzebecher et al. 1988). Excretion proceeded at a ratio of 2:1 with the urine and feces. Cicaprost formed a minor metabolite in man to be found in the urine and accounting for less than 10% of the dose (Hildebrand et al. 1989, 1990). Due to the low dose an isolation and structural elucidation was not possible from human material. Therefore an appropriate animal model was needed.

In whole biosystems of rats, monkeys, and guinea pigs cicaprost was almost stable. Only male rats formed the metabolite, defined as M1, in very low amounts (1%–2% of dose) in a very low urinary dose fraction (3%–5% of dose) after i.v. administration of 0.01 µg [^{3}H]cicaprost/kg. Dogs exhibited a completely different metabolic pattern with a large number of degradation products. However, M1 could not be clearly attributed. Therefore dogs were not only ruled out from further pharmacokinetic studies but also as an animal model in pharma- and toxicology. Liver perfusion models showed no degradation (Hildebrand

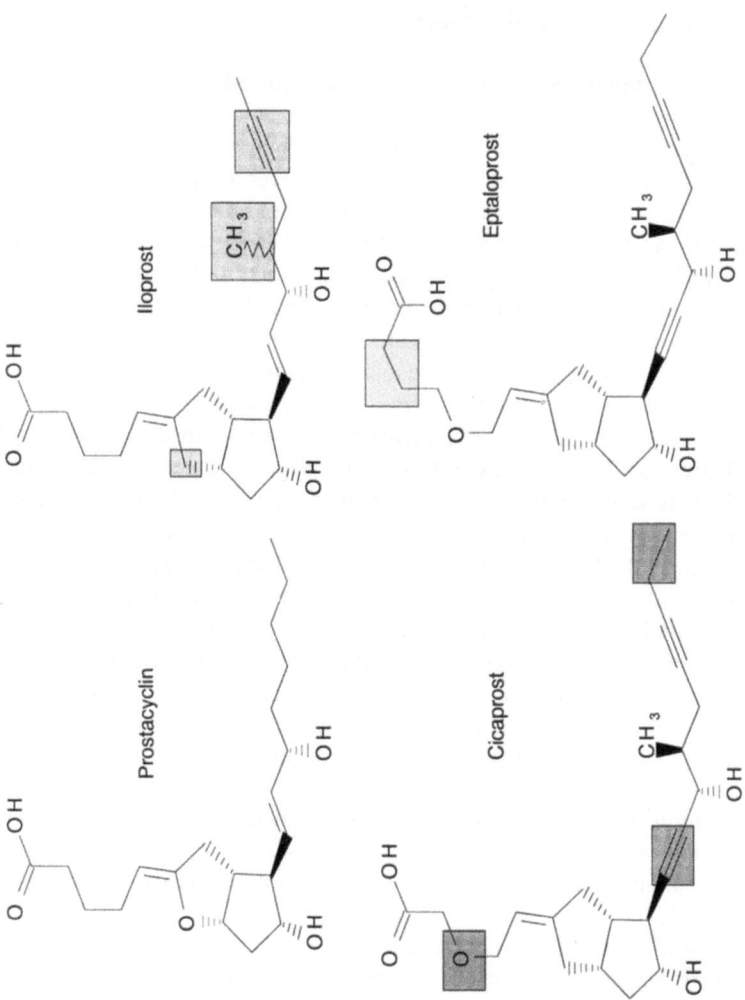

Fig. 1. Prostacyclin and its metabolically stabilized analogues

1986). Freshly prepared rat hepatocytes were originally not considered a successful alternative because a female donor had been used. With the change of sex to male, rat hepatocytes were found to be metabolically active. A further quantitative improvement was achieved by increasing incubation time beyond the range normally used in our laboratory to 3 h. Metabolite fractions increased to an amount that made the hepatocyte model suitable as a factory for M1 (Fig. 2).

To complete the developmental cascade illustrated on Fig. 1 it should be mentioned that eptaloprost was designed as a prodrug of cicaprost to be bioactivated by simple β-oxidation, and that this drug design worked in vitro and in vivo (Hildebrand 1993).

1.2.2 Metabolic Profiling of Ergoline Derivatives

Two structurally highly related ergolines that are subject to extensive biotransformation were investigated by different models. In vivo studies in intact biosystems of rat, dog, and man with ^{14}C-labeled lisuride revealed demethylation at N-6 and various oxidations at different positions of the ergoline system in addition to the mono- and didesalkylation in the urea moiety. Furthermore, in plasma and urine all kinds of products resulting from consecutive metabolic reactions were found. Quantitatively important were the conjugates of ring hydroxylated products and mono- and didesethylated products in rats (Toda and Oshino 1981). In dogs, monkeys, and guinea pigs didesethyllisuride was the main product in urine and feces – similarly to man, where additionally 2-keto-3-OH-lisuride is a main metabolite (Fig. 3; Hümpel et al. 1984). Concerning terguride principally similar reactions were observed. In rats monodesethylterguride was a main excretion form, together with 2-keto-3-OH-terguride (Krause and Hümpel 1988).

Liver perfusion experiments performed primarily in rats and guinea pigs showed a slightly different metabolic pattern. Mono- and didesethyllisuride were the main biodegradation products and C-2 oxidation and N-6 demethylation were observed (Toda and Oshino 1981). A large number of minor polar fractions were seen but not identified. In the case of terguride mono- and didesethyl metabolites were found in rats together with a 2-keto product and a 2-keto-3-OH derivative. In guinea pigs monodesethylterguride was the major radiolabel fraction (Hümpel et al. 1989).

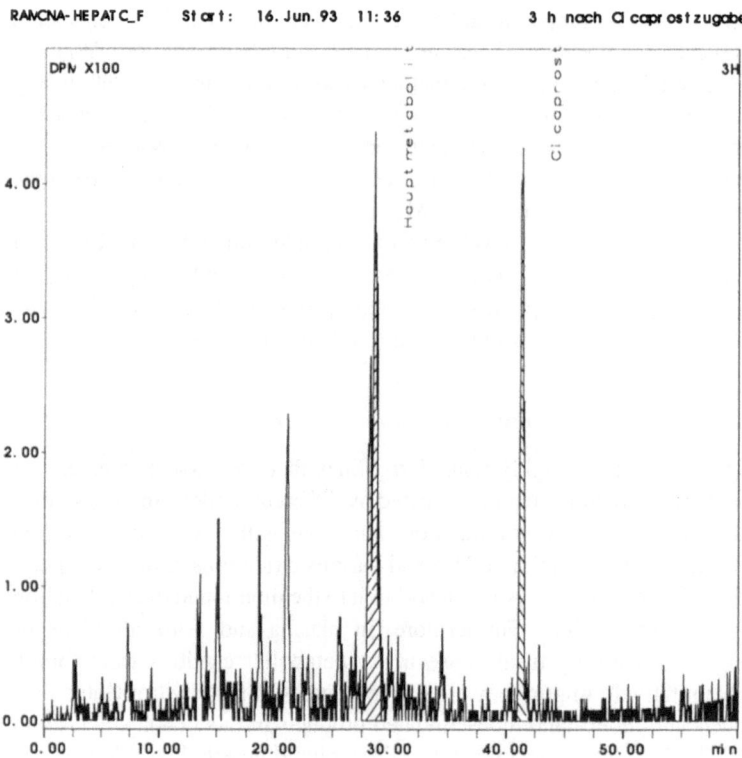

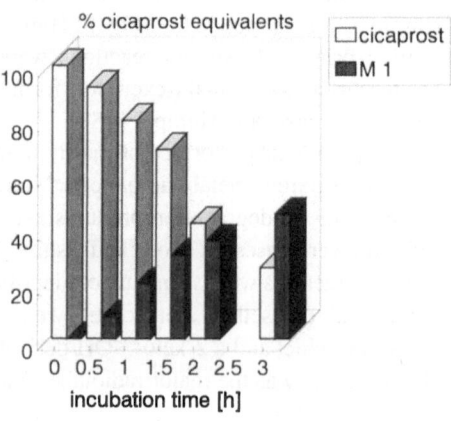

In hepatocytes from various animal species lisuride was transformed to a transient monodesethyl derivative which was subsequently didesethylated in rats and guinea pigs. Some highly polar fractions were seen. In dogs conversion was similar but slower whereas monkeys formed monodesethyllisuride as main product. Terguride was monodesethylated only in hepatocytes from rats and monkeys; in guinea pigs the cascade proceeded further. In hepatocytes terguride was more stable than lisuride (Hümpel et al. 1989).

Experiments with single cytochrome P450 enzyme were performed with both compounds in V79 cells containing rat and human P450 1A1 and 1A2 (Gieschen et al. 1994). Interestingly, both 1A1 enzymes were able to monodesethylate lisuride. This metabolite was not subject to further desalkylation as tested with the reference compound. In the case of terguride monodesethylation was catalyzed only by human 1A1 and was also stopped at the monodesethyl derivative. The corresponding rat enzyme was not able to perform this pathway. The time course of degradation is displayed in Fig. 4. In all experiments the metabolic conversion of the ergolines accounted for a maximum of only 25%, and the metabolites found were in the range of 5%–20%. The most pronounced effect was obtained with human 1A1 in the case of lisuride whereas the monodesethylation of terguride by human 1A1 was also much lower. Translating these data into enzyme activities, the conversion rate was approx. 10 pmol mg^{-1} min^{-1} for rat 1A1 and 20 pmol mg^{-1} min^{-1} for human 1A1; with terguride it dropped to 2 pmol mg^{-1} min^{-1}.

1.2.3 V79 Cell Lines with Single P450 Enzymes as Screening Tool

With regard to single enzymes in permanent cell lines our work up to now has concentrated mainly on using the V79 cell lines (Platt et al. 1989) as a screening tool for substrate properties (Gieschen et al. 1994; Hildebrand et al. 1993). A complete picture is available for only five enzymes, i.e., rat 1A1, 1A2, and 2B1 and human 1A1 and 1A2. We have selected some steroid compounds, the already mentioned ergoline deri-

---

Fig. 2. Use of freshly prepared male rat hepatocytes as metabolite factory to study biodegradation of cicaprost. *Top*, HPLC radiochromatogram from hepatocyte culture; *bottom*, time course of biodegradation

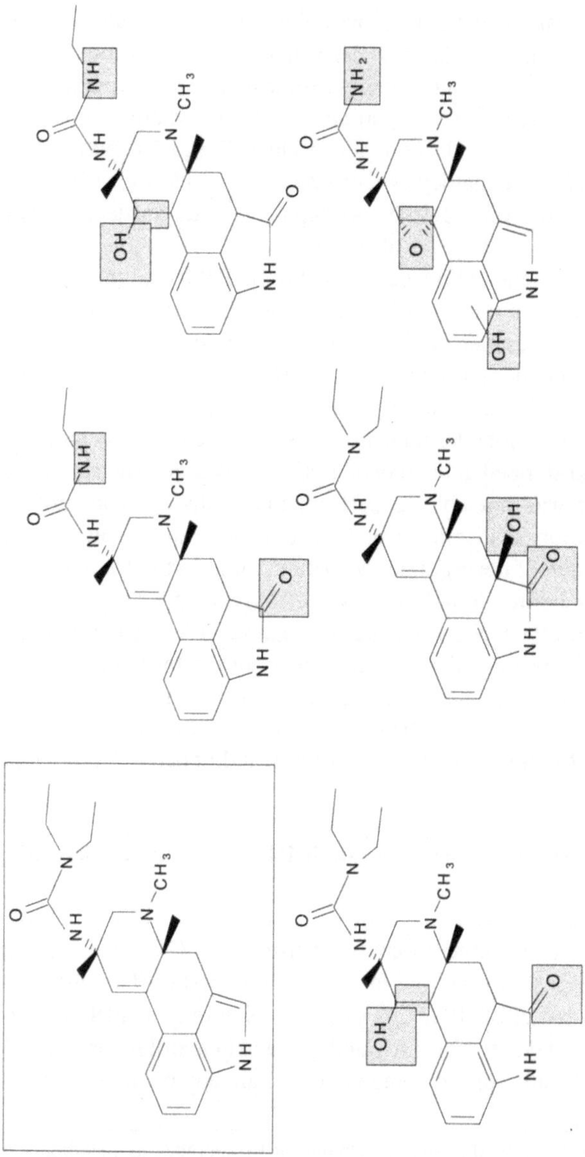

Fig. 3. Selected metabolites of lisuride in intact biosystems

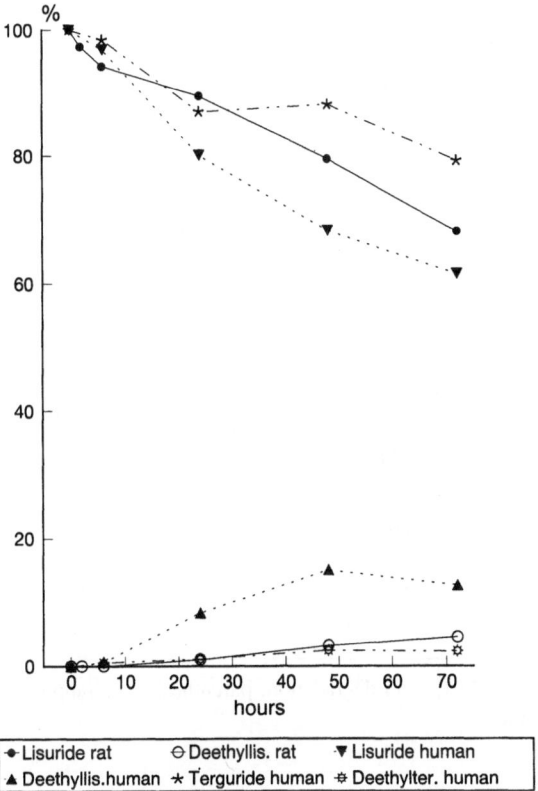

Fig. 4. Time course of the metabolic degradation of lisuride and terguride in V79 cell lines expressing rat and human 1A1

vatives and two β-carbolines and phenimide. The results are summarized in Table 1.

With estradiol, apart from some metabolic activity to form estron which was also present in parent V79 cells, 2-hydroxylation followed by methylation was observed. Ethinylestradiol was not metabolized. Testosterone was used only for the development of methods in the case of 2B1. Neither levonorgestrel or gestodene was subject to any metabolic conversion. Desogestrel was metabolized to the 3-keto derivative by V79 cells, and additionally one or two more polar metabolites were seen

Table 1. Substrate screening with V79 cell lines expressing different P450 enzymes from rat and man

Drug	1A1 (R)	1A2 (R)	2B1 (R)	1A2 (H)	1A1 (H)
Estradiol	(E1)	(E1), 2-Methoxy	–	(E1), 2-Methoxy	(E1), 2-Methoxy
Ethinylestradiol	–	–	–	–	–
Testosterone			16α,β-OH		
Levonorgestrel	–	–	–	–	–
Testodene	–	–	–	–	–
Desogestrel	(3-Keto), +		+		(3-Keto), +(2)
3-Ketodeso-gestrel	+		+	(+)	+
Abecarnil	6-OH	–	–	–	6-OH
Gedocarnil	–	–	–		–
Lisuride	MDL		–	–	MDL
MDL					–
Terguride	(+)	+	–	(+)	MDT
Phenimide				+	+

+ Unknown product; – no degradation; parentheses overlapping activity with parent cell line

with 1A1, similarly to 3-ketodesogestrel. With abecarnil, a main in vivo metabolite, the 6-OH derivative was formed by 1A1 of both species. However, slight structural modifications to gedocarnil resulted in no metabolic activity. In the case of a new developmental compound our cell lines expressing human 1A1 and 1A2 were able to completely degrade phenimide within 24 h, suggesting a low metabolic stability.

Interestingly, V79 cells seem to exhibit relevant phase II activity. Apart from some minor derivatives, a highly polar fraction was detectable after incubation of tritiated estradiol with untransformed V79 cells which was hydrolyzed enzymatically by β-glucuronidase/arylsulfatase. The polar fraction vanished in favor of several products in the retention area of estradiol (Fig. 5). Another interesting result was obtained after incubation of estradiol with V79 cells expressing human 1A2 yielding a

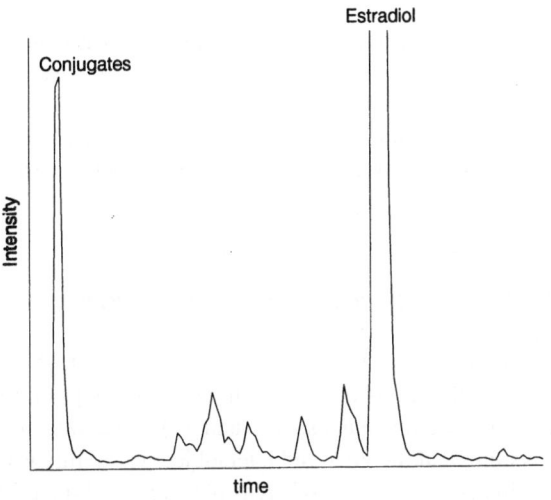

control

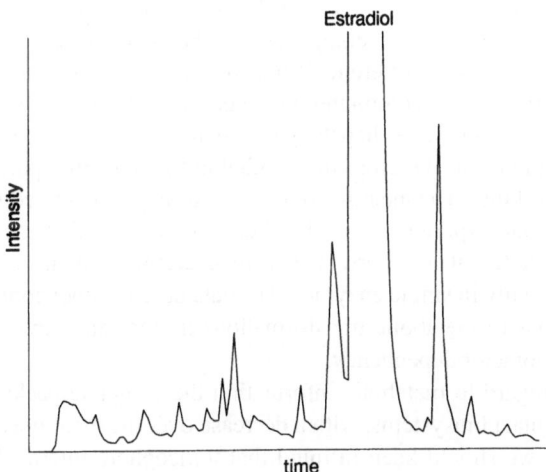

control
after conjugate-cleavage
(Glucuronidase, Arylsulfatase)

Fig. 5. Formation of phase II metabolites by V79 parent cells after incubation with [³H]estradiol

clearcut peak for 2-methoxyestradiol, a compound that must have been generated by methylation after C-2 hydroxylation, an unexpected metabolic activity which was not seen in control cells.

Polar fractions which were hydrolyzed enzymatically have also been observed with other test compounds, and a remarkable phase II activity must therefore be present in V79 cell lines.

1.3 Evaluative Comparison of Different Models

Comparison of the various tools can evaluate the advantages and disadvantages of the different models according to two groups of criteria: (a) the model and experimental aspects and (b) metabolic information (Fig. 6). Physiological integrity and intersubject variability is addressed optimally in intact biosystems. Whereas physiological integrity rapidly decreases when perfused organs or hepatocytes are used, intersubject variability is still reflected by these models. A critical point is animal experimentation. This is required in all cases apart from single enzymes; however, the number of compounds to be investigated in a single experiment is rather different. While intact biosystems and perfused organs fulfill the one-compound/one-experiment criterion, this does not hold true for liver slices, hepatocytes, microsomes, or single enzymes. These methods are therefore more suitable for screening purposes but need some kind of standardization to make experiments comparable. Based on our experience and the literature we feel that the point of required additional cofactors, such as regenerating systems, is a critical point especially in single enzymes. The data base is rather confusing for comparisons of metabolic transformations in different expression systems and cofactor dependency.

With regard to metabolic information the complete package is obtained in intact biosystems, with a decrease over the mentioned models. However, we should keep in mind that toxicophore intermediates are often of high importance, but are often difficult to find in complete biosystems. A similar pattern of complexity of phase I and phase II reactions and all their possible combinations is obtained only in vivo and to lesser extent in the various in vitro models. The largest advantage of intact biosystems is the relationship to other pharmacokinetic processes, such as absorption, distribution, and excretion, which are not

Model aspects	intact biosytem	perfused organ	liver slices	hepato- cytes	micro- somes	single enzyme
physiological integrity						
inter-subject variability						
animal experiment						
one compound/experiment						

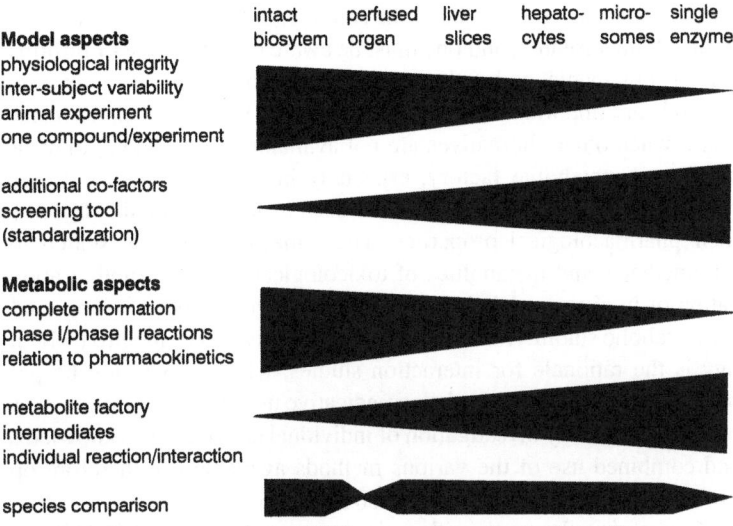

additional co-factors
screening tool
(standardization)

Metabolic aspects
complete information
phase I/phase II reactions
relation to pharmacokinetics

metabolite factory
intermediates
individual reaction/interaction

species comparison

Fig. 6. Evaluative comparison of different tools for metabolic studies

given in all other in vitro models but may be relevant in interpretating data.

The conversely is the case with the formation of metabolites. This task can be achieved by all in vitro models; however, our personal experience focuses on perfusion and hepatocyte models requiring larger amounts. The question of intermediates can best be addressed by microsomes or single enzymes as well as the enzyme kinetics and interactions concerning a special biotransformation pathway. For species comparison intact biosystems may give the most relevant information but, as mentioned above, are not available in early development. Substitutes can be hepatocytes and especially microsomes which can be stored or kept in cultures.

1.4 Discussion and Conclusion

This contribution demonstrates that in vitro approaches will never completely replaced in vivo studies. An in vitro/in vivo correlation is not mandated by any law of nature but has – based upon experiments – be verified on a case-by-case basis. In vitro methods play an important role in metabolic research, and one must be aware of which type of information is obtainable under the experimental conditions used. In vitro methods are important for species comparison in an early experimental phase when other alternatives are not available. Furthermore, they can serve as a metabolite factory, especially in highly metabolized compounds and/or drugs that cannot be dosed as needed in animals due to their pharmacological properties. These may also help to search for intermediates and metabolites of toxicological interest. Another application of in vitro models is in screening a large number of compounds for metabolic stability. One main field of application of in vitro technology is the rationale for interaction studies, which should not be performed based merely upon the comedicative use of another drug but can be substantiated by investigation of individual enzymes. The knowledge and combined use of the various methods available can therefore optimize and facilitate metabolic research.

Concerning the work with a single enzyme expressed in V79 cell lines the inherent metabolic activity of the expression system beyond the transformed P450 characteristics is crucial. The question of model standardization and validation is also important, both for researchers to facilitate comparison of results and for the broader use of such data in discussions with health authorities. Another critical point is the cofactorial demand, which differs from enzyme to enzyme and within one enzyme from substrate to substrate. Lastly, two points are must be considered: the qualitative and quantitative comparison of results with the in vivo situation and the interspecies prediction.

Various models can be used to translate these conclusions to the different phases of drug development (Table 2). In the early phase, when screening and profiling of compounds are of substantial importance, in vitro models can address questions such as metabolic stability, enzyme interactions; and help in species selection in both pharmaco- and toxicology.

During preclinical characterization the metabolic pattern and the structural elucidation of metabolites is of primary interest and might encourage the use of intact animals or perfusion models.

With regard to clinical development we must first validate our animal species selection by comparing metabolic patterns in animals and man. This may give rise to a second-research cycle when a specific

Table 2. Applicability of different metabolic research tools in drug research and development

R + D-Phase	Models	Aims
Screening	Hepatocytes Microsomes Liver slices Single enzymes	Metabolic stability Enzyme interactions Species selection for toxi- cology/pharmacology
Preclinical	Intact animal Liver perfusion	Complete metabolic pattern Isolation of metabolites
Clinical	Man Single enzymes	Metabolic pattern Rationale for interactions Pharmacogenetics

metabolite is to be isolated. An interesting aspect in developing optimal drug therapy is the investigation of possible interactions with other drugs. Here especially single enzymes can help. Additionally, the known polymorphic metabolism pathways can easily be identified in terms of their contribution to the metabolism of a new drug entity. Although in vitro experiments may provide a rationale for in vivo studies, we must be very careful in extrapolating from in vitro results to in vivo situations. Biotransformation is a complex phenomenon, one in which a single enzyme inhibition in vitro may not mean anything at all in vivo due to the availability of alternative pathways.

References

Barr J, Weir AJ, Brendel K, Sipes IG (1991) Liver slices in dynamic organ culture. I. An alternative in vitro technique for the study of rat hepatic metabolism. Xenobiotica 21:331–339

Cholerton S, Daly AK, Idle JR (1992) The role of individual human cytochromes P450 in drug metabolism and clinical response. TIPS 131:434–439

Coundouris JA, Grant MH, Engeset J, Petrie JC, Hawksworth (1993) Cryopreservation of human adult hepatocytes for use in drug metabolism and toxicity studies. Xenobiotica 23:1399–1409

Crespi CL (1991) Expression of cytochrome P450 cDNAs in human β-lymphoblastoid cells: application to toxicology and metabolite analysis. Methods Enzymol 206:123–130

Doehmer J, Oesch F (1991) V79 Chinese hamster cells genetically engineered for stable expression of cytochromes P450. Methods Enzymol 206:117–123

Gieschen H, Hildebrand M, Salomon B (1994) Metabolism of two dopaminergic ergot derivatives in genetically engineered V79 cells expressing CYP450 enzymes. In: Lechner MC (ed) Cytochrome P450 biochemistry, biophysics and molecular biology. John Libbey Eurotext, Paris, pp 467–470

Gonzalez FJ, Idle JR (1994) Pharmacogenetic phenotyping and genotyping – present status and future potential. Clin Pharmacokinet 26:59–70

Gonzalez FJ, Aoyama T, Gelboin HV (1991a) Expression of mammalian cytochrome P450 using vaccinia virus. Methods Enzymol 206:85–92

Gonzalez FJ, Kimura S, Tamura S, Gelboin HV (1991b) Expression of mammalian cytochrome P450 using baculovirus. Methods Enzymol 206:93–99

Guillouzo A, Begue JM, Campion JP, Gascoin MN, Gugen-Guillouzo C (1985) Human hepatocyte culture; a model of pharmaco-toxicological studies. Xenobiotica 15:635–641

Hildebrand M (1986) Studies on the pharmacokinetics of ZK 96 480, a novel PGI_2-mimetic, in rat and cynomolgus monkey. Prostaglandins 32:425–438

Hildebrand M (1992) Pharmacokinetics of iloprost and cicaprost in mice. Prostaglandins 44:431–442

Hildebrand M (1993) Bioactivation of eptaloprost in animals and man. Prostaglandins 46:177–189

Hildebrand M, Schütt A, Staks T, Matthes H (1989) Pharmacokinetics of ^{3}H-cicaprost in healthy volunteers. Prostaglandins 37:259–272

Hildebrand M, Staks T, Nieuweboer B (1990) Pharmacokinetics and pharmacodynamics of cicaprost in healthy volunteers after oral administration of 5 to 20 g. Eur J Clin Pharmacol 39:149–153

Hildebrand M, Gieschen H, Salomon B (1993) Metabolism of selected sex steroids by rat CYP1A1, 1A2, and 2B1 expressed in V79 cell lines. In: Lechner MC (ed) Cytochrome P450 biochemistry, biophysics and molecular biology. John Libbey Eurotext, Paris, pp 693–696

Hümpel M, Krause W, Hoyer GA, Wendt H, Pommerenke G (1984) The pharmacokinetics and biotransformation of ^{14}C-lisuride hydrogen maleate in rhesus monkey and man. Eur J Drug Metab Pharmacokin 9:347–357

Hümpel M, Sostarek D, Gieschen H, Labitzky C (1989) Studies on the biotransformation of lonazolac, bromerguride, lisuride and terguride in laboratory animals and their hepatocytes. Xenobiotica 19:361–377

Krause W, Hümpel M (1988) Pharmacokinetics of the dopamine partial agonist, terguride, in the rat and the rhesus monkey. Eur J Drug Metab Pharmacokin 13:185–194

Meijer KF, Keulemans K, Mulder G (1981) Isolated perfused rat liver technique. Methods Enzymol 77:81–93

Platt KL, Molitor E, Döhmer J, Dogra S, Oesch F (1989) Genetically engineered V79 chinese hamster cell expression of purified cytochrome P450IIB1 monooxygenase activity. J Biochem Toxicol 4:1–6

Stürzebecher S, Hildebrand M, Schöbel C, Seifert W, Staks T (1988) Platelet inhibitory and hemodynamic effects of a new stable PGI$_2$ analogue, cicaprost (ZK 96 480), in different animal species and in man. Biomed Biochim Acta 47:45–47

Toda T, Oshino N (1981) Biotransformation of lisuride in the hemoglobin free perfused rat liver and in the whole animal. Drug Metab Dispos 9:108–113

2 Regulation of Xenobiotic-Metabolizing Cytochromes P450

F. J. Gonzalez

2.1   Introduction   21
2.1.1   History and Structure.   21
2.1.2   Nomenclature.   23
2.1.3   Evolution   24
2.1.4   Human P450s.   27
2.2   Regulation   27
2.2.1   Receptor-Mediated Regulation   28
2.2.2   Tissue-Specific Regulation   31
2.2.3   Posttranscriptional Regulation.   34
2.2.4   Clinical Consequence of P450 Regulation   36
References   37

2.1 Introduction

2.1.1 History and Structure

Cytochromes P450 were identified as chromophores in the late 1950s (Garfinkle 1958; Klingenberg 1958) and later purified, characterized, and named "cytochrome P450" on the basis of a Soret absorbance band when reduced and complexed with carbon monoxide (Omura and Sato 1961, 1964a,b). The latter property is due to the presence of a noncovalently bound heme in the form of protoporphyrin IX.

   With the exception of bacterial P450s, all P450s are membrane bound. A few P450s involved in steroid biosynthesis are found in the

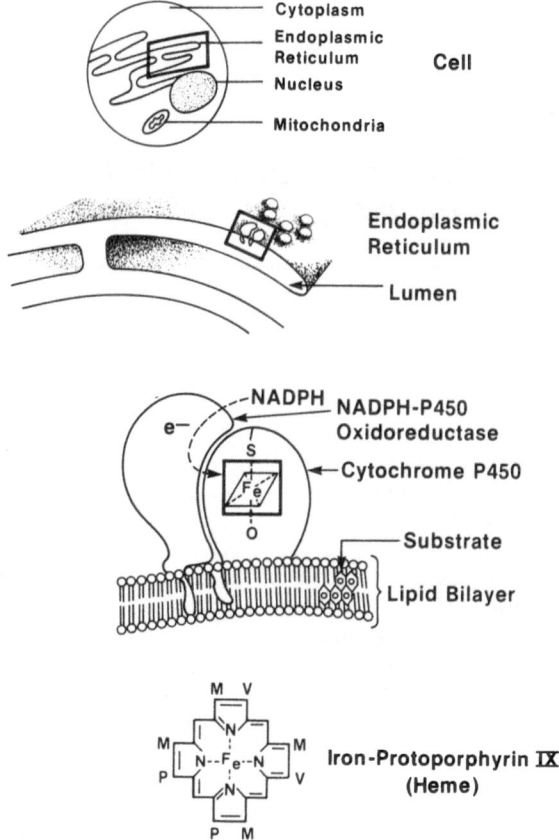

Fig. 1. Schematic representation of the microsomal P450

inner mitochondrial membrane while most P450s are in the bilayer of
the endoplasmic reticulum (Fig. 1). It is believed that the hydrophobic
amino terminal "signal sequence" directs the P450 into the lipid bilayer
during synthesis (Ahn et al. 1993), and that about 20%–30% of the
protein is imbedded in the membrane, with the bulk of the enzyme
facing the cytoplasmic surface. This membrane localization is ideally
suited to the functional role of P450s to oxidize foreign and hydrophobic chemicals that can diffuse into the intracellular membrane net-

work of the cell. Eukaryote P450s usually contain 500 ± 20 amino acids. They all possess a highly conserved segment of amino acids near the carboxy terminus that is centered around a Cys residue (Nebert and Gonzalez 1987) that donates the thiolate ligand to the heme iron (White and Coon 1980). This region has been used to isolate P450 cDNAs from cultured cells (Shen et al. 1993) and petunia flowers (Holten et al. 1993).

2.1.2 Nomenclature

The primary amino acid sequences of the P450s are the basis for the P450 nomenclature systems (Nebert et al. 1987; Nelson et al. 1993). All P450s are members of the P450 superfamily which is subdivided into families and subfamilies. By comparing the sequence relatedness of all P450s, without consideration of their functions or substrate specificity, percentage boundaries were established to delineate families and subfamilies. With the latest nomenclature update in which 221 sequences were compared, 36 families were identified, 12 of which are found in mammals. A P450 within one family is usually defined as having at least 40% amino acid sequence similarity to a P450 protein in another family. Only a few exceptions to this rule have been found (Nelson et al. 1993). Within a given subfamily in mammals P450s are usually more than 55% identical. In some cases confusion results from comparison of distantly related species such as birds and fish. In these cases P450s having suspected evolutionary relationships to certain mammalian P450 families are incorporated as subfamilies even though they exhibit as low as 46% sequence relatedness (Nelson et al. 1993). P450s are named with the root CYP (all capital letters) followed by an Arabic number designating a family, a capital letter designating the subfamily, and another Arabic number denoting the individual P450 form. Two problems exist with the present nomenclature system. First, it is difficult to distinguish whether two highly similar P450s from the same species are derived from separate genes or are allelic variants. This determination can be made only when detailed analysis of gene locus structures are made. Unless the genes are particularly important, they are not usually analyzed. An example of an allelic variant of a mouse gene giving rise to two almost identical P450s having different catalytic activities has been discussed (Gonzalez 1992a).

A second problem with the nomenclature system is that it is not easy to determine species-orthologous counterparts (derived from the same immediate ancestor gene) in complex subfamilies. The problem becomes more severe when P450s from distantly related species are compared. This problem can be illustrated by analysis of the CYP2D subfamilies from humans, rats, and mice (Gonzalez and Nebert 1990). As a result, in the more complex subfamilies the individual P450 forms have been assigned numbers based on when they were entered into the nomenclature system. Therefore it is not possible to determine what species a P450 is derived from by simply knowing its name. Even two P450s from different species having similar catalytic activities (e.g., rat CYP2D1 and human CYP2D6) have different names. Another annoying aspect of the nomenclature system is that the numbers designating the P450 forms do not run consecutively for a given species. Thus, when reading the P450 literature, one must be armed with the most recent P450 nomenclature update to determine the species from which a particular P450 was derived. With the exception of the steroidogenic P450s in which the family numbers correspond to their steroid hydroxylation sites, it is not possible to predict catalytic activities on the basis of the name of a P450. All problems considered with perfect hindsight, the nomenclature system for P450s is a vast improvement over the plethora of trivial names from different laboratories that existed prior to 1987.

2.1.3 Evolution

The marked diversity of P450s and species differences in their expression and catalytic activities have led to speculation about the driving forces behind the evolution of these enzymes (Nebert and Gonzalez 1985; Nelson and Strobel 1987; Gonzalez and Nebert 1990). We know only the results of P450 evolution, especially that pertaining to those P450s that metabolize foreign compounds. Evolution has resulted in species differences in number of P450 genes, regulation of P450s, and catalytic activities of individual P450 forms. The study of P450 evolution is not only of immense academic interest but is also of practical importance with respect to human drug metabolism and susceptibility to environmental toxins and carcinogens.

Analysis of the P450 evolutionary tree suggest that the oldest and most well-conserved P450 genes are those that metabolize steroids, particularly the steroid biosynthetic P450s. This could be due to the role of steroids in maintenance of the structures of lipid bilayers (Nebert and Gonzalez 1987). Later, steroid hormones were required for differentiation of multicellular organisms (Nebert 1992). It has also been suggested that early P450s evolved to metabolize toxic atmospheric oxygen and that this became linked to metabolism of other chemicals (R. Feyereisen, personal communication).

The proliferation in number of P450s associated with foreign compound metabolisms appears to have occurred during the last several hundred million years (Nelson and Strobel 1987). Since most of these P450s are not known to have any role in physiologically relevant metabolism of endogenous chemicals such as hormones, they probably evolved to metabolize foreign chemicals. These chemicals would likely be from dietary plant sources and would include endogenous plant toxins that evolved to make them impalantable to animals. The animal-plant "warfare" hypothesis was developed to attempt to explain the evolution of the P450s (Gonzalez and Nebert 1992). In this hypothesis plants developed toxins, or phytoalexins, as a means of surviving in an environment of animal predators. Animals countered by having enzymes such as P450s that could rapidly inactivate the plant toxins. Over a period of millions of years, as plants evolved new toxins and animals evolved variant or new P450s to deal with these toxins, the P450s superfamily would expand and change. This is the most likely explanation for the marked species differences in foreign compound metabolizing P450s that are observed today since each species would be exposed to distinct plant habitats. Indeed, today a number of plant-derived chemicals or their derivatives are used as drugs, and these drugs are inactivated by P450s.

The molecular basis for species differences in P450s is a process called DNA turnover. It is believed that DNA is under constant flux such as gene duplications, unequal crossing over, slippage during replication, insertions, deletions, and gene conversions. These processes can result in formation of new genes or alteration of preexisting genes. In the case of P450s one can imagine that these new or variant P450 genes can be used to degrade a dietary toxin or to allow an animal to use another plant source. The gene can then become fixed in a population. If a P450 gene is no longer needed, it can be dispensed of. This process might be evident in the occurrence of polymorphisms which could be considered as defective genes on the way out. Polymorphisms in P450s are found in rats, mice, rabbits, and humans.

If the above scenario of P450 evolution is correct, humans would no longer have that selective pressure to use their P450s or to develop new P450 activities. Humans can now control their habitat and diet and thus can avoid the exposure to plant toxins. If the xenobiotic P450s have no other crucial function than to degrade plant toxins, the nonessential P450s would ultimately disappear through accumulation of mutations. As discussed below, human P450 polymorphisms have been demonstrated in the CYP2C and CYP2D subfamilies. All other known human xenobiotic metabolizing P450s are expressed albeit at highly variable levels in the population.

Table 1. Major drug-metabolizing P450 enzymes

P450	Enzyme substrate
CYP1A2	Imipramine, caffeine, phenacetin, verapamil
CYP2C8, CYP2C9, CYP2C18	Benzphetamine, diazepam, diclofenac, hexobarbital, ibuprofen, imipramine, oxicam, anti-inflammatory drugs, proguanil, propranolol, retinoic acid, S-warfarin, naproxen, tolbutamide, tetrahydrocannabinol
CYP2C19	S-Mephenytoin, omeprazole
CYP2D6	Antiarrhythmic agents, antihypertensives, β-blockers, monoamine oxidase inhibitors, morphine derivatives, antipsychotics, tricyclic antidepressants
CYP2E1	Chlorzoxazone
CYP3A3, CYP3A4, CYP3A5	Aldrin, benzphetamine, cyclosporin, erythromycin, lidocaine (lignocaine), lovastatin, midazolam, quinidine, 17α-ethynylestradiol, terfenadine, triazolam, various 1,4-dihydropyridines
CYP4A11	Leukotriene receptor antagonist (long-chain fatty acid hydroxylase)

Table 2. Major carcinogen/mutagen-metabolizing P450s

P450	Enzyme substrate
CYP1A1	Polycyclic aromatic hydrocarbons
CYP1A2	Food mutagens, aflatoxins
CYP2A6	Certain nitrosamines
CYP2B6	Cyclophosphamide, ifosfamide
CYP2E	Numerous low molecular weight suspect carcinogens (acrylonitrile, benzene, nitrosamines, vinyl halides)
CYP3A3 CYP3A4 CYP3A5 CYP3A7	Aflatoxins, food mutagens, nitroaromatic hydrocarbons

All CYP3A P450s are believed to have similar substrate specificities

2.1.4 Human P450s

The notable species differences in catalytic activities of P450s has led to the direct analysis of human P450s. A limited number of P450s have been purified from liver tissues, and a larger number have been identified through cDNA cloning (Gonzalez 1992a; Gonzalez and Idle 1994). Immunocorrelative and chemical inhibition/stimulation analyses using human liver banks have led to the identification of preferred drug (Table 1) and carcinogen/mutagen (Table 2) substrates for select P450 forms. It is noteworthy that the drugs are largely metabolized P450s in the CYP2C, CYP2D, and CYP3A subfamilies, whereas the carcinogens and mutagens are metabolized by the CYP1A, CYP2E, and to some extent CYP3A subfamilies.

2.2 Regulation

Enzymes are controlled at both the transcriptional and posttranscriptional levels. P450s are under diverse transcriptional control as determined by a large number of studies carried out in rodents (Gonzalez et

al. 1993). Practically every mammalian P450 has its own regulatory
circuit. Certain P450s are present at low levels and are induced by
foreign compounds while others are constitutively expressed in certain
tissues. Posttranscriptional regulation is less well studied and has been
recognized only for CYP2E1 (Song et al. 1987, 1989).

2.2.1 Receptor-Mediated Regulation

The *Ah-Receptor*. The mechanism of regulation of CYP1A1 has been
extensively studied (Nebert et al. 1991; Swanson and Bradfield 1993;
Whitlock 1993). The *CYP1A1* gene has one or more segments of DNA
upstream of its transcription start sites called Ah-receptor regulatory
elements, or AhREs (Nebert and Jones 1989). Binding of the Ah-receptor to the AhRE activates transcription of the *CYP1A1* gene. The Ah-
receptor, in its active form, actually consists of heterodimeric subunits,
the ligand-binding domain (ALBD) and the Ah-receptor nuclear trans-
porter (ARNT). The proposed mechanism for regulation of CYP1A1 is
shown schematically in Fig. 2. In the absence of ligand the ALBD is
associated with heat shock protein 90 (HSP-90). Upon ligand binding to
the ALBD, the HSP-90 dissociates and ARNT binds, yielding the receptor complex capable of interacting with the AhRE. Other elements,
including a basal transcription element (BTE) also appears to play a role
in maximizing transcription activation through the BTE binding protein
(Fujii-Kuriyama et al. 1992; Imataka et al. 1992; Sogawa et al. 1993),
however, this element has been characterized only in the rat *CYP1A1*
gene. Existence of a homologous region in the corresponding human
and mouse genes has not been demonstrated. A role for protein kinase C
(PKC) and phosphorylation on CYP1A1 induction has been proposed
based on studies showing the effects of phorbol ester (Okino et al. 1992;
Moore et al. 1993) and PKC inhibitors (Carrier et al. 1992). In vitro
studies have demonstrated an effect of phosphorylation on Ah-receptor
binding to AhRE (Pongratz et al. 1991; Carrier et al. 1992). Results
using specific inhibitors of PKC indicate that phosphorylation is not
required for Ah-receptor-specific DNA binding (Schafer et al. 1993).
Thus, the role of Ah-receptor phosphorylation in the induction process
remains unclear.

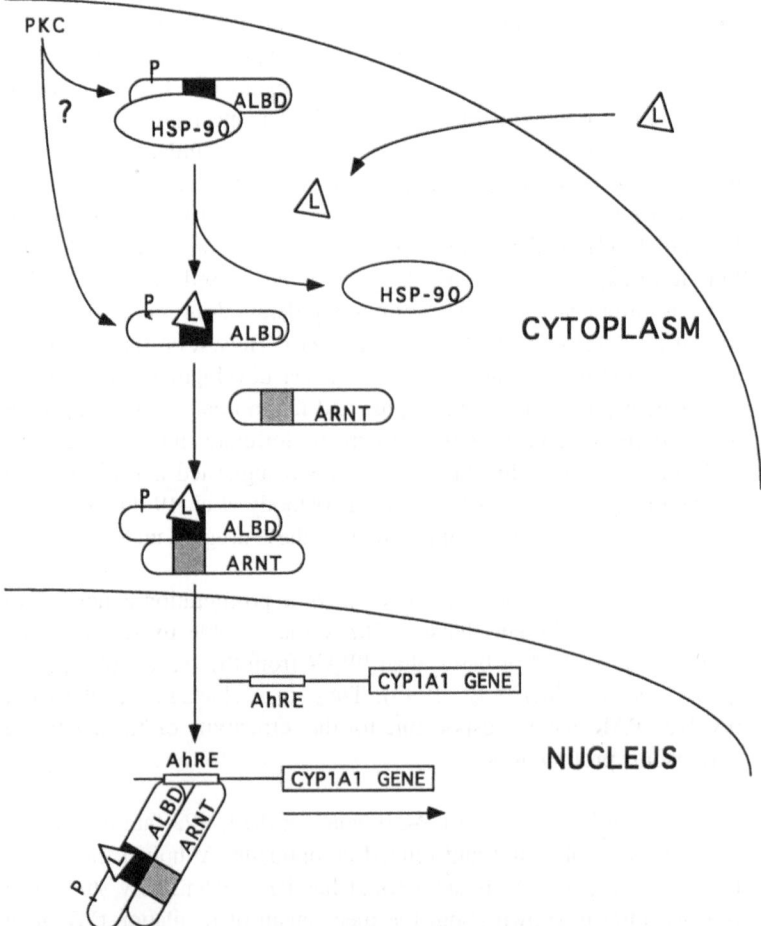

Fig. 2. Mechanism for the transcriptional activation of the *CYP1A1* gene

Peroxisome Proliferator Activated Receptor. CYP4A P450s can be induced by compounds called peroxisome proliferators that include clofibrate (Kimura et al. 1989). This induction is mediated by a family of receptors called the peroxisome proliferator activated receptors (PPAR; Green 1992; Muerhoff et al. 1992). At least five subtypes of PPAR are believed to exist in any given animal (Chen et al. 1993).

PPAR (Issemann and Green 1990; Sher et al. 1993) and PPARγ (Zhu et al. 1993) are able to *trans*-activate gene expression through binding to a PPAR response element located upstream of PPAR-responsive genes (Green 1992; Muerhoff et al. 1992). The mechanism of gene activation of peroxysome proliferators is presumed similar to the steroid hormones. Peroxisome proliferators are known to cause hepatomegaly and induce the number and size of peroxysomes in rodents (Reddy and Lalwani 1983). Rodents are also susceptible to liver cancer following long-term exposure to fibrate drugs. This has raised the concern that humans chronically administered hypolipidemic drugs could be at risk for cancer development (Reddy et al. 1980). Thus, it becomes an issue as to whether humans are at risk for cancer development. It is known that humans, as well as guinea pigs and marmosets, are less sensitive to the peroxisome proliferation effects of clofibrate and related agents, and limited epidemiological evidence have suggested that patients on fibrate drugs are not at risk for cancer (Moody et al. 1991). However, until the mechanism of carcinogeneity of these drugs is understood, the concern remains.

The resistance of humans to peroxisome proliferation is not due to lack of PPAR. Humans have a PPAR that is able to activate gene transcription as well or better than PPAR from the susceptible species mouse and rat (Sher et al. 1993). Thus, a mechanism that does not involve PPAR may be responsible for the refractivity of humans to the peroxisome proliferators.

Phenobarbital. A number of P450 genes in the CYP2 families are induced by phenobarbital and related compounds. Although the induction of P450 genes by phenobarbital has been extensively studied in rodents, little is known about the mechanism of regulation (Waxman and Azaroff 1992). An upstream regulator element has been identified in an inducible chicken P450 gene. However, a phenobarbital receptor has not been identified. Interestingly, phenobarbital also induces a P450 gene in bacteria. The mechanism of induction in bacteria has been identified (Wen et al. 1989; Shaw and Fulco 1992). Activation of the bacterial CYP102 gene is due to release of inhibition of a repressor protein by phenobarbital (Shaw and Fulco 1993).

2.2.2 Tissue-Specific Regulation

Most xenobiotic-metabolizing P450s are expressed in liver, although low level expression is found in extrahepatic tissues such as lung and gastrointestinal tract. Rodents have a number of hepatic P450s that are under developmental and sex-dependent control (Table 3). The sex-dependent expression is due to differences in circulating growth hormone patterns between males and females (Gonzalez 1992b), a phenomenon that is not found in humans or rabbits.

Liver-specific expression of genes has been the most actively studied in the area of tissue-specific gene regulation owing to the abundance and cellular homogeneity of this tissue. Genes encoding serum albumin, transferrin, α_1-antitrypsin, fibrinogen, amylase, glucose 6-phosphate dehydrogenase, and other liver enzymes and proteins have been intensively studied. A number of transcription factors have been identified that are enriched in liver tissue (Table 4).

Several families of factors have been identified that are distinguished based on their conserved DNA-binding domains. The domains homeo and forkhead were first described in *Drosophila*. The zinc finger domain of HNF-4 is similar to that found in the steroid receptor superfamily. The bZIP domain was described in the C/EBP proteins isolated from rat liver. DBP is a member of the C/EBP family but lacks a leucine zipper domain.

The factors listed in Table 4 are enriched in liver; however, they can also be expressed to some degree in extrahepatic tissues. For example, some of the C/EBP factors are found in adipose tissue and lung. It should also be noted that other members of these families exist that may not be expressed in liver. For example, a number of HNF-3-like forkhead homologues have been isolated that are expressed in brain, kidney, lung, and intestine (Clevidence et al. 1993).

It is not unreasonable to assume that the liver-enriched transcription factors that have been characterized using other liver-specific genes are also involved in expression of the P450s. Such has been demonstrated by recent studies described below.

Cytochromes P450 have complex and diverse regulatory control circuits. Certain P450 forms are expressed at very low levels and are induced by foreign compounds such as polycyclic aromatic hydrocarbons, phenobarbital, and clofibrate. Other P450s are constitutively ex-

Table 3. Modes of regulation of rat hepatic cytochrome P450 genes

P450	Expression pattern
CYP2A1 CYP3A2	Activated after birth in both sexes, then repressed in females at puberty
CYP2E1 CYP2D1 CYP2D2 CYP2D5	Activated after birth in both sexes
CYP2C6 CYP2C7 CYP2D3	Activated in both sexes at puberty
CYP2A2 CYP2C11 CYP2C13 CYP2C22 CYP4A2	Activated in males at puberty
CYP2C12	Activated in females at puberty

Table 4. Hepatocyte-enriched transcription factors (from Lai and Darnell 1991; Williams et al. 1991)

Family (DNA binding motif)	Transcription factor
POU homeodomain	HNF-1α (LF-B1, APF) HNF-1β (vHNF-1)
Forkhead	HNF-3α HNF-3β HNF-3γ
Zinc finger	HNF-4
bZIP (basic region, leucine zipper)	C/EBPα C/EBPβ C/EBPγ DBP

HNF, Hepatocyte nuclear factor; *C/EBP*, CCAAT-enhancer binding protein; *CRB*, *C/EBP*-related protein; *DBP*, albumin D site binding protein

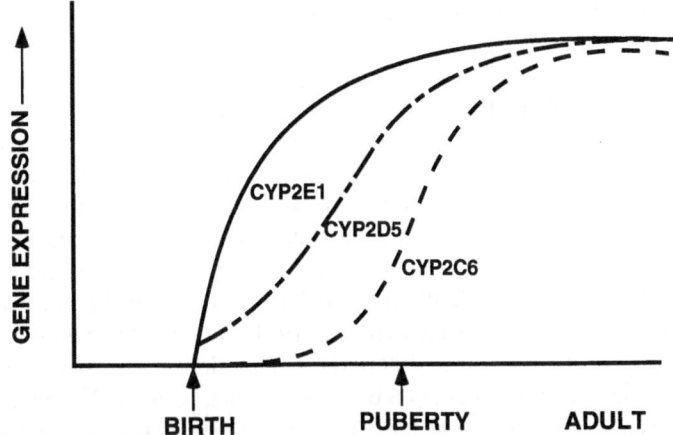

Fig. 3. Transcriptional activation of the rat *CYP2E1*, *CYP2D5*, and *CYP2C6* genes during development

pressed. These enzymes are enriched in liver tissues, and their expression is activated during development. The rat has been used as an experimental model to investigate the mechanisms of tissue-specific, developmentally programmed transcription of P450s in the CYP2 family. Developmental activation of the *CYP2E1*, *CYP2D5*, and *CYP2C6* genes are shown in Fig. 3. A variety of experimental approaches have been used including in vitro transcription, heterologous promoter transcription assays, *trans*-activation transcription assays, and in vitro DNA binding assays including gel mobility shift, DNase I footprinting, and methylation interference to determine their mechanisms of regulation. These studies have shown that the *CYP2C6* promoter is controlled by the liver-enriched transcription factor DBP (Yano et al. 1992). The C/EBPα and C/EBPβ proteins are capable of binding the regulatory region, but they cannot activate the *CYP2C6* promoter. *CYP2E1* is under control of the factor HNF-1α (Ueno and Gonzalez 1990) and *CYP2D5* is regulated in part by Sp1 and C/EBPβ (Lee et al. 1994). These data indicate that the developmentally activated expression of P450 genes in rat liver is due in part to the presence and activity of liver-enriched transcription factors.

2.2.3 Posttranscriptional Regulation

Some P450s can be posttranslationally regulated by their own substrates. A number of cases exist in which P450s are inhibited. This inhibition can either be competitive or noncompetitive and irreversible (Guengerich and Shimada 1991; Guengerich et al. 1991). For example, the antibiotic triacetyoleandomycin inhibits CYP3A4. This noncompetitive inhibition is due to the binding of its metabolite to the heme group of the P450 and is highly specific for CYP3A4 (Guengerich and Shimada 1991; Chang et al. 1994). CYP2E1 is strongly inhibited by diethyldithiocarbomate (Guengerich et al. 1991) although the inhibition is less specific (Chang et al. 1994). This inhibition is also mechanism based and is due to production of an active metabolite. α-Naphthoflavone is a potent inhibitor of the CYP1A1 and to an extent other P450s, and CYP1A2 is strongly inhibited by furaphylline (Sesardic et al. 1990) and fluvoxamine (Brosen et al. 1993).

P450s can also be posttranscriptionally activated. The activity of CYP3A4 and other CYP3A enzymes are increased in the presence of certain chemicals such as 7,8-benzoflavone and even CYP3A substrates (Schwab et al. 1988; Johnson et al. 1988; Guengerich and Shimada 1991). This unusual type of activation is due to the presence of either two substrates or a substrate and activator at the enzymes active site (Shou et al. 1994). Thus, 7,8-benzoflavone, when present at the active site, would allow faster metabolism of substrates such as testosterone. This might be due to the closing of the active site space allowing for exclusion of water molecules that could accelerate substrate uncoupling or a more efficient environment for oxygen activation (Fig. 4, scheme 1). Others (Schwab et al. 1988; Johnson et al. 1988) have proposed an allosteric mechanism (Fig. 4, scheme 2).

P450s can also be increased in level through stabilization mechanisms. CYP2E1 protein is stabilized by some of its own substrates. This is due to an elimination of a rapid phase of protein degradation (Song et al. 1989). CYP2E1 is degraded in a biphasic manner with half-lives of 7 and 37 h. Acetone, a CYP2E1 substrate, causes a loss of the 7-h component, resulting in an increase in CYP2E1 protein. The mechanism for this stabilization effect may involve phosphorylation (Fig. 5). CYP2E1 substrates inhibit phosphorylation of Ser-129, and this phosphorylation event is thought to trigger degradation via an ATP-depend-

SCHEME 1   SCHEME 2

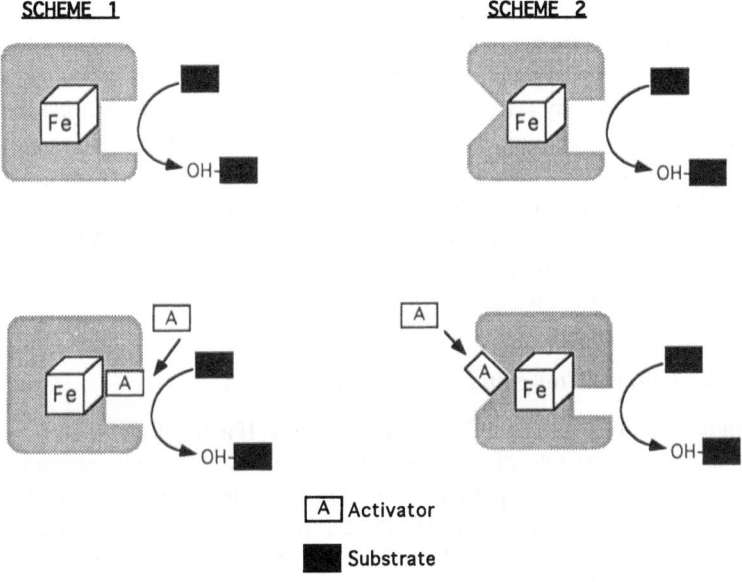

A Activator

■ Substrate

Fig. 4. Activation of P450s. *Scheme 1*, the effect of the activator at the P450 active site; *scheme 2*, a typical allosteric activation

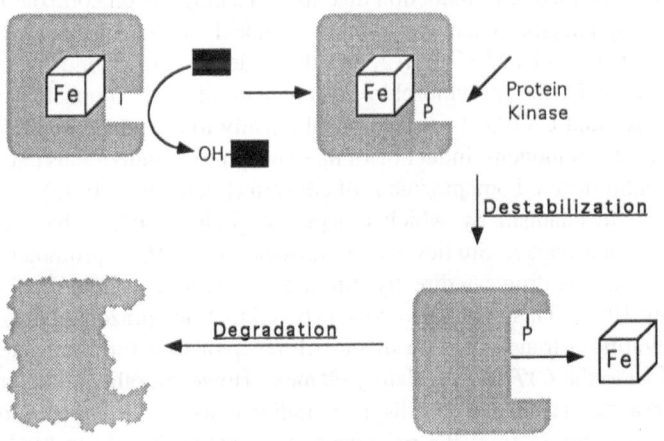

Fig. 5. Mechanism for substrate-induced stabilization of CYP2E1

ent microsomal protease (Eliasson et al. 1990, 1992). Two serine proteases purified from microsomal membranes are capable of degrading
CYP2E1 modified by phosphorylation (Zhukov et al. 1993). Modification of CYP2E1 by activated substrates such as carbon tetrachloride can
also trigger proteolytic degradation (Tierney et al. 1992). Hormones
such as glucagon are thought to increase the phosphorylation of
CYP2E1 and stimulate its turnover (Eliasson et al. 1992).

After exposure to its substrates, the mRNA encoding CYP2E1 is
also markedly stabilized (Song et al. 1987). The mechanism of this
stabilization effect is not currently known.

2.2.4 Clinical Consequence of P450 Regulation

Induction of P450s can influence drug therapy. If a drug induces its own
metabolism, the rate of clearance of the drug, if chronically administered, increases over a period of time. Omeprazole, a selective inhibitor
of the H+/K+-adenosine triphosphatase proton pump, was shown to
induce CYP1A1 and CYP1A2 in liver (Diaz et al. 1990) and CYP1A1
in the gut (McDonnell et al. 1992). Interestingly, this drug is oxidatively
metabolized by a P450 in the CYP2C family, the S-mephenytoin 4'-hy-
droxylase, which is expressed polymorphically in humans (Rost et al.
1992; Chiba et al. 1993). Thus subjects able to metabolize omeprazole
are more refractive to induction than those lacking the omeprazole-me-
tabolizing enzyme. There are no known clinical consequences of induction of CYP1A1 and CYP1A2, and the wide therapeutic index of the
drug renders the polymorphic metabolism of little impact. Since
CYP1A1 and CYP1A2 are able metabolically to activate a number of
chemical carcinogens, induction of these enzymes in individuals chronically administered omeprazole is of concern (Lucier et al. 1992).

The mechanism by which omeprazole induces P450s has come
under controversy. Studies using transfected *CYP1A1* promoter revealed that the drug can directly stimulate transcription (Quattrochi and
Tukey 1993). This study also demonstrated that omeprazole is capable
of eliciting a transformation of the Ah-receptor to a form capable of
binding to the *CYP1A1* regulatory element. However, others found that
omeprazole is unable to displace radiolabeled 2,3,7,8-tetrachloro-
dibenzo-*p*-dioxin from the receptor and is unable directly to bind the
receptor (Daujat et al. 1992).

A more serious consequence of P450 regulation is the inhibition of P450 enzymes. A large number of antibiotics can inhibit their own metabolism and the metabolism of other therapeutically used drugs (Gillum et al. 1993). This is due to the enzyme CYP3A4 which is able to metabolize a large number of drugs (Table 1). CYP2C9 metabolism and clearance of drugs such as warfarin is also potently inhibited by sulfaphenazole (Rettie et al. 1992). The serious consequence of P450 inhibitions has led to widespread use of cDNA-expressed P450s to establish the P450 form responsible for metabolism of a candidate drug during early stages of its preclinical development (Gonzalez 1992c).

References

Ahn K, Szezesna-Skorupa E, Kemper B (1993) The amino-terminal 29 amino acids of cytochrome P450 2C1 are sufficient for retention in the endoplasmic reticulum. J Biol Chem 268:18726–18733

Brosen K, Skjelbo E, Rasmussen BB, Poulsen HE, Loft S (1993) Fluroxamine is a potent inhibitor of cytochrome P4501A2. Biochem Pharmacol 45:1211–1214

Carrier F, Owens RA, Nebert DW, Puga A (1992) Dioxin-dependent activation of murine Cyp1a-1 gene transcription requires protein kinase C-dependent phosphorylation. Mol Cell Biol 12:1856–1863

Chang TKH, Gonzalez FJ, Waxman DJ (1994) Evaluation of triacetyloleandomycin, α-naphtoflavone and diethyldithiocarbamate as selective chemical probes for inhibition of human cytochrome P450 (submitted)

Chen F, Law SW, O'Malley BW (1993) Identification of two mPPAR related receptors and evidence for the existence of five subfamily members. Biochem Biophys Res Commun 196:671–677

Chiba K, Kobayashi K, Manabe K, Tuni M, Kamataki T, Ishizaki T (1993) Oxidative metabolism of omeprazole in human liver microsomes: cosegregation with S-mephenytoin 4'-hydroxylation. J Pharmacol Exp Ther 266: 52–59

Clevidence DE, Overdier DG, Tao W, Qian X, Pani L, Lai E, Costa RH (1993) Identification of nine tissue-specific transcription factors of the hepatocyte nuclear factor 3 forkhead DNA-binding domain family. Proc Natl Acad Sci USA 90:3948–3952

Daujat M, Peryt B, Lesca P, Fourtanier G, Domergue J, Maurel P (1992) Omeprazole, an inducer of human CYP1A1 and 1A2, is not a ligand for the Ah receptor. Biochem Biophys Res Commun 188:820–825

38   F. J. Gonzalez

Diaz D, Fabre I, Daujat M, Saint Aubert B, Bories P, Michel H, Maurel P
(1990) Omeprazole is an aryl hydrocarbon-like inducer of human hepatic
cytochrome P450. Gastroenterology 99:737–747

Eliasson E, Johansson I, Ingelman-Sundberg M (1990) Substrate-, hormone-
and cAMP-regulated cytochrome P450 degradation. Proc Natl Acad Sci
USA 87:3225–3229

Eliasson E, Mkrtchian S, Ingelman-Sundberg M (1992) Hormone- and substrate-regulated intracellular degradation of cytochrome P450 (2E1) involving MgATP-activated rapid proteolysis in the endoplasmic reticulum. J
Biol Chem 267:15765–15769

Fujii-Kuriyama Y, Imataka H, Sogawa K, Yasumoto K, Kikuchi Y (1992)
Regulation of CYP1A1 expression. FASEB J 6:706–710

Garfinkle D (1958) Studies on pig liver microsomes. I. Enzyme and pigment
composition of different microsomal fractions. Arch Biochem Biophys
77:493–509

Gillum JG, Israel DS, Polk RE (1993) Pharmacokinetic drug interactions with
antimicrobial agents. Clin Pharmacokinet 25:450–482

Gonzalez FJ (1992a) Control of constitutively-expressed developmentally-ac-
tivated rat hepatic cytochrome P450 genes. Keio J Med 41:68–75

Gonzalez FJ (1992b) Human P450s: problems and prospects. Trends Pharmacol Sci 13:346–352

Gonzalez FJ (1992c) In vitro systems for prediction of rates of drug clearance
and drug-drug interactions. Anesthesiology 77:413–415

Gonzalez FJ, Idle JR (1994) Pharmacogenetic phenotyping and genotyping:
present status and future potential. Clin Pharmacokinet 26:59–70

Gonzalez FJ, Nebert DW (1990) Evolution of the P450 gene superfamily: animal plant warfare, molecular drive and human genetic differences in drug
oxidation. Trends Genet 6:182–186

Gonzalez FJ, Liu SY, Yano M (1993) Regulation of cytochrome P450 genes:
molecular mechanisms. Pharmacogenetics 3:51–57

Green S (1992) Receptor-mediated mechanism of peroxysome proliferators.
Biochem Pharmacol 43:393–401

Guengerich FP, Shimada T (1991) Oxidation of toxic and carcinogenic chemicals by human cytochrome P450 enzymes. Chem Res Toxicol 4:391–407

Guengerich FP, Kim DH, Iwasaki M (1991) Role of human cytochromes P-
450 IIE1 in the oxidation of many low molecular weight cancer suspects.
Chem Res Toxicol 4:168–179

Holton TA, Brugliera F, Lester DR, Tanaka Y, Hyland CD, Menting JGT, Lu
CY, Farcy E, Stevenson TW, Cornish EC (1993) Cloning and expression
of cytochrome P450 genes controlling flower colour. Nature 366:276–279

Imataka H, Sogawa K, Yasumoto K, Kikuchi Y, Sasano K, Kobayashi A, Hayami M, Fujii-Kuriyama Y (1992) Two regulatory proteins that bind to the basic transcription element (BTE), a GC box sequence in the promoter region of rat P-4501A1 gene. EMBO J 11:3663–3671

Issemann I, Green S (1990) Activation of a member of the steroid hormone receptor superfamily by peroxysome proliferators. Nature 347:645–650

Johnson EF, Schwab GE, Vickery LE (1988) Positive effectors of the binding of an active site-directed steroid to rabbit cytochrome P450 3c. J Biol Chem 263:17672–17677

Kimura S, Hardwick JP, Kozak CA, Gonzalez FJ (1989) The rat clofibrate-inducible CYP4A subfamily II. cDNA sequence of IVA3, mapping of the CYP4a locus to mouse chrosome 4, and coordinate and tissue-specific regulation of the CYP4A genes. DNA 8:517–526

Klingenberg M (1958) Pigments of rat liver microsomes. Arch Biochem Biophys 75:376–386

Lai E, Darnell JE (1991) Transcriptional control in hepatocytes: a window on development. Trends Biol Sci 16:427–430

Lee Y-H, Yano M, Liu S-Y, Matsunaga E, Johnson PF, Gonzalez FJ (1994) A novel cis-acting element controlling the rat CYP2D5 gene requiring cooperativity between C/EBP and an Sp1 factor. Mol Cell Biol 14:1383–1394

Lucier GW, Thompson CL, Hoel DG (1992) Omeprazole, cytochrome P450 and chemical carcinogenesis. Gastroenterology 103:1509–1516

McDonnell WM, Scheiman JM, Traber PG (1992) Induction of cytochrome P4501A genes (CYP1A) by omeprazole in the human alimentary tract. Gastroenterology 103:1509–1516

Moody DE, Reddy JK, Lake BG, Popp JA, Reese DH (1991) Peroxysome proliferation and nongenotoxic carcinogens: commentary on a symposium. Fundam Appl Toxicol 16:233–248

Moore M, Narasimhan TR, Sternburg MA, Wang X, Safe S (1993) Potentiation of CYP1A1 gene expression in MCF-7 human breast cancer cells cotreated with 2,3,7,8-tetrachlorodibenzo-p-dioxin and 12-O-tetradecanoylphorbol-13-acetate. Arch Biochem Biophys 305:483–488

Muerhoff AS, Griffin KJ, Johnson ER (1992) The peroxisome proliferator-activated receptor mediates the induction of CYP4A6, a cytochrome P450 fatty acid-hydroxylase, by clofibrate. J Biol Chem 267:19051–19053

Nebert DW (1992) Proposed role of drug-metabolizing enzymes: regulation of steady state levels of the ligands that effect growth homeostasis, differentiation, and neuroendocrine functions. Mol Endocrinol 5:1203–1214

Nebert DW, Gonzalez FJ (1987) P450 genes: structure, evolution and regulation. Annu Rev Biochem 56:943–993

Nebert DW, Gonzalez FJ (1985) Cytochrome P450 gene expression and regulation. Trends Pharmacol Sci 6:160–164

Nebert DW, Jones JE (1989) Regulation of the mammalian cytochrome P450 (CYP1A1) gene. Int J Biochem 21:243–252

Nebert DW, Nelson DR, Coon MJ, Estabrook RW, Gonzalez FJ, Guengerich FP, Gunsalus IC, Johnson EF, Kemper B, Levin W, Phillips IR, Waterman MR (1987) The P-450 gene superfamily: recommended nomenclature. DNA 6:1–11

Nebert DW, Petersen DD, Puga A (1991) Human AH locus polymorphism and cancer: inducibility of CYP1A1 and other genes by combustion products and dioxin. Pharmacogenetics 1:68–78

Nelson DR, Strobel HW (1987) Evolution of cytochrome P450 proteins. Mol Biol Evol 4:572–593

Nelson DR, Kamataki T, Waxman DJ, Guengerich FP, Estabrook RW, Feyereisen R, Gonzalez FJ, Coon MJ, Gunsalus IC, Gotoh O, Okuda K, Nebert DW (1993) The P450 superfamily: update on new sequences, gene mapping, accession numbers, early trivial names of enzymes, and nomenclature. DNA Cell Biol 12:1–51

Okino S, Pendurthi UR, Tukey RH (1992) Phorbol esters inhibit the dioxin receptor-mediated activation of mouse Cyp1a-1 and Cyp1a-2 gene by 2,3,7,8-tetrachlorodibenzo-p-dioxin. J Biol Chem 267:6991–6998

Omura T, Sato R (1961) A new cytochrome in rat liver microsomes. J Biol Chem 237:1375–1376

Omura T, Sato R (1964a) The carbon monoxide-binding pigment of liver microsomes. I. Evidence for its hemoprotein nature. J Biol Chem 239:23702378

Omura T, Sato R (1964b) The carbon monoxide binding pigment of liver microsomes. II. Solubilization, purification, and properties. J Biol Chem 239:2379–2385

Pongratz I, Stromstedt PE, Mason GG, Pollinger L (1991) Inhibition of the specific DNA binding activity of the dioxin receptor by phosphatase treatment. J Biol Chem 266:16813–16817

Quattrochi LC, Tukey RH (1993) Nuclear uptake of the Ah (dioxin) receptor in response to omeprazole: transcriptional activation of the human CYP1A1 gene. Molec Pharmacol 43:504–508

Reddy JK, Lalwani ND (1983) Carcinogenesis by hepatic peroxisome proliferators: evaluation of the risk of hypolipidemic drugs and industrial plasticizers to humans. Crit Rev Toxicol 12:1–58

Reddy JK, Azarnoff DL, Hignite CE (1980) Hypolipidaemic hepatic peroxisome proliferators form a novel class of chemical carcinogens. Nature 283:397–398

Rettie, AE, Korzekwa KR, Kunze KL, Lawrence RF, Eddy AC, Aoyama T, Gelboin HV, Gonzalez FJ, Trager Wf (1992) Hydroxylation of warfarin by human cDNA-expressed cytochrome P-450: a role for P-4502C9 in the etiology of (S)-warfarin-drug interactions. Chem Res Toxicol 5:54–59

Rost KL, Brosicke H, Brockmoller J, Scheffler M, Helge H, Roots I (1992) Increase of cytochrome P4501A2 activity by omeprazole: evidence by the ^{13}C-[N-3-methyl]-caffeine breath test in poor and extensive metabolizers of S-mephenytoin. Clin Pharmacol Ther 52:170–180

Schafer MW, Madhukar BV, Swanson HI, Tullis K, Denison MS (1993) Protein kinase C is not involved in Ah receptor transformation and DNA binding. Arch Biochem Biophys 307:267–271

Schwab GE, Rauch JL, Johnson EF (1988) Modulation of rabbit and human hepatic cytochrome P-450-catalyzed steroid hydroxylations by α-naphthoflavone. Mol Pharmacol 33:493–499

Sesardic D, Boobis AR, Murray BP, Murray S, Segura J, Torre RDL, Davis DS (1990) Furaphylline is a potent selective inhibitor of cytochrome P4501A2 in man. Br J Clin Pharmacol 29:651–663

Shaw GC, Fulco AJ (1992) Barbiturate-mediated regulation of expression of the cytochrome P450BM-3 gene of Bacillus negatarium by Bm3R1 protein. J Biol Chem 267:5515–5526

Shaw GC, Fulco AJ (1993) Inhibition by barbiturates of the binding of Bm3R1 repressor to its operator site on the barbiturate-inducible cytochrome P450BM-3 gene of Bacillus negatarium. J Biol Chem 268:2997–3004

Shen Z, Wells RL, Liu J, Elkind MM (1993) Identification of a cytochrome P450 by reverse transcription-PCR using degenerate primers containing inosine. Proc Natl Acad Sic USA 90:11483–11487

Sher T, Yi HF, McBride OW, Gonzalez FJ (1993) cDNA cloning, chromosomal mapping and functional characterization of the human peroxisome proliferator activated receptor. Biochemistry 32:5598–5604

Shou M, Grogan H, Mancewicz JA, Krausz KW, Gonzalez FJ, Gelboin HV, Korzehwa KR (1994) Activation of CYP3A4: evidence for the simultaneous binding of two substrates in a cytochrome P450 active site (submitted)

Sogawa K, Imataka H, Yamasaki Y, Kusume H, Abe H, Fujii-Kuriyama Y (1993) cDNA cloning and transcriptional properties of a novel GC box-binding protein, BTEB2. Nucl Acids Res 21:1527–1532

Song BJ, Matsunaga T, Hardwick JP, Veech RL, Yang CS, Gelboin HV, Gonzalez FJ (1987) Stabilization of P450j mRNA in the diabetic rat. Mol Endocrinol 1:542–547

Song BJ, Veech RL, Park SS, Gelboin HV, Gonzalez FJ (1989) Induction of rat hepatic N-nitrosodimethylamine demethylase by acetone is due to protein stabilization. J Biol Chem 264:3568–3572

Swanson HI, Bradfield CA (1993) The AH-receptor: genetics, structure and function. Pharmacogenetics 3:213–231

Tierney DJ, Hass AL, Koop DR (1992) Degradation of cytochrome P450 2E1: selective loss after labelization of the enzyme. Arch Biochem Biophys 293:9–16

Ueno T, Gonzalez FJ (1990) Transcriptional control of the rat hepatic CYP2E1 gene. Mol Cell Biol 10:4495–4505

Waxman DJ, Arzaroff L (1992) Phenobarbital induction of cytochrome P-450 gene expression. Biochem J 281:577–592

Waxman DJ, Hansen AJ, May BK (1991) Transcriptional regulation of a phenobarbital-responsive enhancer domain. J Biol Chem 266:17031–17039

Wen LP, Ruettinger RR, Fulco AJ (1989) Requirement for a 1'-kilobase 5'-flanking sequence for barbiturate-inducible expression of the cytochrome P-450BM3 gene in Bacillus negatarium. J Biol Chem 264:10996–11003

White RE, Coon MJ (1980) Oxygen activation by cytochrome P450. Annu Rev Biochem 49:315–356

Whitlock JP (1993) Mechanistic aspects of dioxin action. Chem Res Toxicol 6:754–763

Williams SC, Cantwell CA, Johnson PF (1991) A family of C/EBP-related proteins capable of forming covalently linked leucine zipper dimers in vitro. Genes Dev 5:1553–1567

Yano M, Falvey E, Gonzalez FJ (1992) Role of the liver-enriched transcription factor DBP in expression of the cytochrome P450 CYP2C6 gene. Mol Cell Biol 12:2847–2854

Zhukov A, Werlinder V, Ingelman-Sundberg M (1993) Purification and characterization of two membrane bound serine proteinases from rat liver microsomes active in degradation of cytochrome P450. Biochem Biophys Res Commun 197:221–228

Zhu Y, Alveres K, Huang Q, Rao MS, Reddy JK (1993) Cloning of a new member of the peroxisome proliferator-activated receptor gene family from mouse liver. J Biol Chem 268:26817–26820

3 Cytochrome P450 in Human Drug Metabolism: How Much Is Predictable?

U. A. Meyer

3.1	Introduction	43
3.2	Variability of Cytochrome P450 Function	44
3.2.1	Induction	44
3.2.2	Inhibition	45
3.2.3	Disease	45
3.2.4	Genetic Polymorphism	45
3.3	How Much Is Predictable?	49
3.4	Human Liver Bank	50
3.5	Specific Probes and Tools To Study Human Cytochrome P450	51
3.6	In Vitro Prediction of Metabolic Pathways	52
3.7	Limitations of In Vitro Approaches	53
3.8	Examples	54
3.9	Outlook	54
References		54

3.1 Introduction

Cytochrome P450 (CYP) mono-oxygenases represent one of the major enzyme systems that determine the organism's capability of dealing with drugs and chemicals. Studies over the past 20 years have provided evidence that cytochromes P450 occur in many different forms (isoforms or isozymes) which differ in spectral, chemical, and immunological properties and have different substrate affinities. These isozymes also differ in their regulation and tissue distribution. Recombinant DNA studies indicate that between 50 and 200 structural genes may code for different cytochrome P450 isozymes in a single organism. Close to 30 human cytochrome P450 genes have now been characterized. The multiplicity of P450 isozymes explains in part the literally thousands of substrates known to be metabolized by this system (for review, see Nelson et al. 1993).

3.2 Variability of Cytochrome P450 Function

Numerous factors influence the concentration and activity of cytochromes P450, including age, nutrition, liver disease, environmental chemicals (e.g., cigarette smoking), and drugs themselves. A major source of interindividual differences in drug metabolism are common genetic polymorphisms, i.e., inherited variations in single cytochrome P450 enzymes and other drug metabolizing-enzymes (Meyer et al. 1990). These variants include "no function", "decreased function", and even "increased function" alleles or haplotypes of the respective genes (for review, see Meyer 1994). Recent studies also indicate that there are major interethnic differences in drug metabolism, particularly in the expression of cytochromes P450 and acetyltransferases, including markedly different frequencies of genetic polymorphisms (Kalow et al. 1986). Most P450 isozymes are regulated by multiple mechanisms, but transcriptional activation of P450 genes is the predominant mechanism (Gonzalez 1992).

3.2.1 Induction

A large number of P450 enzymes (and also other drug metabolizing enzymes such as glucuronosyltransferases and epoxide hydrolase) are selectively induced by drugs such as phenobarbital, rifampicin, dexamethasone, glutheimide, phenacetin, phenytoin, and antipyrine or by cigarette smoke (Okey 1990; Waxman and Azaroff 1992). The induction response is dose dependent and reversible. The exact mechanism of this induction response is still unknown, except for the induction of certain P450s by polycyclic aromatic hydrocarbons such as dioxin via the Ah-receptor in mice (Whitlock 1993). Induction leads to progressive drug tolerance and many clinically significant drug-drug interactions. There are considerable species and strain differences in response to inducers. Extrapolation of the inducing potential of a new compound from animal studies to man is still very difficult (Okey 1990; Waxman and Azaroff 1992).

3.2.2 Inhibition

Since one P450 enzyme may bind many different substrates, competition at the site of drug binding or at other sites of the reaction cycle by cosubstrates or other chemicals may cause inhibition. Known potent inhibitors of high specificity of P450 enzymes are quinidine (CYP2D6), cimetidine (numerous P450s), furafylline (CYP1A2), and gestodene (CYP3A subfamily). A summary of these inhibitors is presented in Table 1.

3.2.3 Disease

Severe liver disease (hepatitis, cirrhosis, liver cancer), diabetes, and starvation may affect cytochrome P450 by multiple mechanisms. Of particular importance are drugs whose metabolism is limited by liver blood flow (e.g., propropolol, lidocaine, nifedipine, verapamil). In cirrhosis with portal-systemic shunts these drugs can reach very high (toxic) plasma concentrations because the portal blood bypasses the liver, preventing metabolism of the drug.

3.2.4 Genetic Polymorphism

Genetic variation in drug metabolizing enzymes can markedly affect drug kinetics (Meyer et al. 1990; Kalow 1992). Several of these genetic polymorphisms are now well studied at the epidemiological, protein, and DNA levels, and simple DNA-tests to genotype individuals have been developed (Table 2).

Table 1. Typical substrates and inhibitors of human cytochromes P450

Cytochrome P450 isoforms	Model substrates	Inhibitors	Regulation of isoform activity in vivo
CYP1A1, CYP1A2	Caffeine (N^3-demethylation), phenacetin	Ellipticine, α-naphtoflavone furafylline	Induced in cigarette smokers
CYP2A6	Coumarin		
CYP2C9, CYP2C10	Tolbutamide, phenytoin	Sulfaphenazole	Induced by rifampicin; rare genetic defect
CYP2C19	S-Mephenytoin, omeprazole, proguanil, diazepam	Tranylcypromine	Genetic polymorphism
CYP2D6	Debrisoquine, dextromethorphan, metoprolol, sparteine, bufuralol	Quinidine	Genetic polymorphism
CYP2E1	Chlorzoxazone, ethanol, N-nitrosodimethylamine, 4-nitrophenol	Diethyldithiocarbamate	Induced be ethanol
CYP3A3, CYP3A4, CYP3A5	caffeine (8-hydroxylation), cyclosporin, dihydropyridines, lidocaine, midazolam	Troleoandomycin (and other macrolides), gestodene, α-naphthoflavone	Induced by dexamethanose, rifampicine, phenobarbital 3A5 polymorphically expressed (present in ~20% of livers)

Table 2. Genetic polymorphisms of cytochrome P450

Gene/enzyme	Discovery	Substrates	Incidence of "slow" metabolizer	Significance
CYP1A1	Genetic difference in inducibility, RFLP	Polycyclic aromatic hydrocarbons	?	Risk factor for cancer in smokers (?)
CYP1A2	Caffeine metabolisms	Arylamines, nitrosamines polycyclic aromatic hydrocarbons	10%	Variability in drug response
CYP2C19	Mephenytoin polymorphism	Mephenytoin, omeprazole, proguanil, etc.	2%–5% Caucasians 15%–20% Asians	Variability in drug and carcinogen effects
CYP2D6	Debrisoquine/sparteine polymorphism	Antidepressants, betaadrenergic blockers, antiarrhythmics, neuroleptics, codein, etc.	5%–10% Caucasians 0%–2% Asians	Variability in drug response
CYP2E1	RFLPs, altered regulation	Nitrosamines, ethanol, chlozoxazone, etc	?	Risk factor for liver cirrhosis?

The *debrisoquine/sparteine-type polymorphism* of drug oxidation affects the metabolism of over 30 clinically used drugs; these include the following:

- Antiarrhythmic agents
    - Encainide
    - Flecainide
    - Mexiletine
    - Perhexiline
    - Propafenone
    - *N*-Propylajmaline
    - Sparteine
- Antidepressants
    - (+)-Amiflamine
    - Amitryptiline
    - Clomipramine despiramine
    - Imipramine
    - Nortriptyline
- Antipsychotics
    - Perphenazine
    - Thioridazine
    - Clozapine
- β-Adrenoceptor blocking agents
    - Alprenolol
    - Bopindolol
    - Bufuralol
    - Metoprolol
    - Penbutolol
    - Propranolol
    - Timolol
- Others
    - Codeine
    - Debrisoquine
    - Dextromethorphan
    - Guanoxan
    - Indoramin
    - 4-Methoxyamphetamine
    - Phenformin
    - Tropisetron

A cytochrome P450 (CYP2D6) is deficient in "poor metabolizers" of debrisoquine (Zanger et al. 1988). In leukocyte DNA of poor metabolizer subjects and their families mutant alleles of the CYP2D gene associated with the PM phenotype were identified and sequenced (for review, see Heim and Meyer 1990; Broly et al. 1991; Meyer 1994). Two point mutations and a deletion of the entire CYP2D6 gene (CYP2D6-A, -B, -D) account for over 75% of the mutations causing deficient CYP2D6 function in Caucasians. DNA restriction fragment length analysis and a simple DNA test using polymerase chain reaction amplification can identify or predict more than 90% of poor metabolizers (Broly et al. 1991). More recently, CYP2D6 alleles with only slightly decreased function (CYP2D6-C, -F, -G, -J, -Ch1) in "intermediate" metabolizers as well as duplications of a functional CYP2D6 gene (CYP2D6-L) or amplification to up to 12 CYP2D6 genes on chromosome 22 have been observed in so-called ultrarapid metabolizers (Johansson et al. 1993). Thus, numerous CYP2D6 alleles with normal, decreased, absent, or increased function now explain the large interindividual variation in the metabolic ratio in extensive and poor metabolizers of debrisoquine.

The *mephenytoin-type polymorphism* affects the metabolism of mephenytoin and several other drugs. It is caused by a deficiency of CYP2C19 (Goldstein et al. 1994). Several important clinical drugs are involved in this polymorphism, including omeprazol, proguanil, and diazepam (Wilkinson et al. 1992). A molecular defect in CYP2C19 has recently been discovered and accounts for approx. 75% of mutant alleles in Caucasians and Asians (DeMorais et al. 1994). The clinical importance of this polymorphism has been less well studied. DNA polymorphisms have also been described for human CYP1A1, 2A6, and CYP2E1. In some studies but not in others the CYP1A1 and CYP2E12E1 DNA polymorphisms have been associated with increased individual risk for certain cancers (for review, see Daly et al. 1993).

3.3 How Much Is Predictable?

In the context of the considerable advances in our understanding of human cytochromes P450 and the increasing availability of probes and tools to study these enzymes, the question is whether we can use this

knowledge to obtain reliable estimates of human disposition of a drug
before its first administration to man, and whether important genetic and
environmental variation including drug interactions can be predicted
from studies in animals or studies in vitro. The questions that one may
ask at the various stages of drug development include the following:

- Preclinical
  - Can we predict human drug metabolism?
  - From animal data?
  - From in vitro studies?
  - From computer models?
- Phase I studies
  - Do they confirm in vitro data?
  - Do they help in risk assessment?
- Clinical drug development
  - How well can we predict kinetics?
  - How well can we predict environmental and genetic variation?

The objectives of in vitro drug metabolism studies are: (a) to assist in the
drug discovery process; (b) to assist in the preclinical evaluation of
drugs; and (c) to develop the methodology for clinical studies.

   The following describes the experimental approaches to assessing
human drug metabolism including many cytochrome P450s that have
been used in my laboratory and in many others to achieve these goals.

3.4 Human Liver Bank

We have used a large human liver bank to study drug-metabolizing
enzymes. Two types of tissue are collected: organ transplant donor
livers and wedge biopsies taken during abdominal surgery. Donor livers
are used for large-scale purification of enzymes, the purified proteins
serving to raise antibodies. From the same livers mRNA is prepared and
cDNA expression libraries are constructed. cDNAs coding for a specific
drug-metabolizing enzyme are isolated with antibodies, with oligonu-
cleotides constructed on the basis of amino acid sequence information
or with cDNAs for the corresponding animal enzymes. The cDNAs are
then used to isolate the gene and to determine its chromosomal localization and its structure. Both human liver tissue and subcellular fractions

such as microsomes are now commercially available in most countries. An important aspect of this is that human cytochromes P450 enzymes and their mRNAs are well preserved in frozen human liver for many years. Wedge biopsies from patients undergoing abdominal surgery are important for in vivo/in vitro comparisons (Meier et al. 1983)

At present our liver bank has over 50 human livers, most of which have been characterized for several drug-metabolizing enzyme activities, and in which the content of a particular enzyme hase been estimated by V_{max} and western blot analysis. Our bank includes livers of "poor metabolizers" of the genetic polymorphisms of the debrisoquine, mephenytoin, and acetylation types. For the most common enzymes we have developed chemical, immunological, and cDNA probes or have received these by exchange with other laboratories or commercially. The enzymes include CYP1A1, 1A2, 2C8/9/10, 2C19, 2D6, 2E1, 3A4, 3A5 as well as epoxide hydrolase and two N-acetyltransferases (NAT1 and NAT2). We have expressed all of these enzymes in heterologous systems such as COS-1 cells.

3.5 Specific Probes and Tools To Study Human Cytochromes P450

The increasing availability of specific antibodies and selective substrates and inhibitors for each major human P450 enzyme provide the tools for initial studies in subcellular fractions from liver. Another recent advance in this field is the development of more efficient analytical methods such as liquid-phase ionization techniques combined with MS/MS analysis. These approaches allow extremely rapid and sensitive identification and quantification of metabolites. Finally, for most human P450s enzymes expressed from cloned cDNAs are commercially available either as microsomal fractions from heterologous expression systems or as cell lines with a unique, permanently expressed gene for a drug-metabolizing enzyme. These are discussed in detail in other contributions to this volume.

3.6 In Vitro Prediction of Metabolic Pathways

In vitro incubation of a new compound with microsomes, 9000 g supernatants, or of liver slices or other subcellular fractions can identify the metabolic pathways and the relative intrinsic clearances (V_{max}/K_m) of each pathway, but this requires that the metabolites have been identified and can be analyzed. Once the major metabolic pathways are known, the contribution of drug-metabolizing enzymes for each pathway can in principle be determined, by experiments often called "reaction phenotyping."

Competitive Inhibition. At an early phase of its study a new compound can be added to microsomal assays of model substrates for specific P450-isozymes or to an assay of a single enzyme in a heterologous expression system. If competitive inhibition of the model reaction occurs, the affinity with which the new compound interacts with the inhibited isoform (K_i) can be determined. This P450 isozyme is a good a candidate to catalyze the metabolism of the new compound. However, competitive inhibition per se does not prove that this isoform metabolizes the compound in question. For example, quinidine is a potent competitive inhibitor of CYP2D6, but is metabolized mostly by CYP3A4 (Guengerich et al. 1986). On the other hand, if the K_i is in the range of concentration likely to occur in the liver during therapy with the new compound, drug-drug interactions with other substrates of this enzyme are likely. Quinidine indeed predictably inhibits the metabolism of many other substrates of CYP2D6. This approach has the big advantage that it can be used before assays for the metabolism of the new compound are available, and that large numbers of compounds can be screened (see for instance Fonné-Pfister and Meyer 1988).

Antibodies and Selective Chemical Inhibitors. Once an assay for the reaction in question is available, a panel of inhibitory antibodies can be used to identify the enzyme(s) responsible for the production of a particular metabolite. However, these are relatively difficult experiments and the reproducibility of the degree of inhibition of an enzyme by monoclonal or polyclonal antibodies is relatively poor. An alternative is the correlation of an activity with the amount of enzyme recognized on immunoblots using isoform-specific antibodies. We regularly use a panel of livers for each of which the quantity of several P450 isozymes have been determined by immunoblotting. This does not require inhibitory antibodies, and western blots give more reproducible results than antibody inhibition experiments. More recently, a number of selective chemical inhibitors have become available for the major drug-metabolizing cytochrome P450s (Table 1). Chemical inhibitors are excellent tools for testing the involvement of a particular isozyme and also allow reliable results of the extent of the contribution of this isozyme to a particular metabolic pathway. Finally, performing the assay in microsomes or cells with single expressed cytochromes P450 identifies which P450 can perform the metabolic reaction and with what type of affinity. However, this approach does not indicate how important this enzyme is for the overall metabolism of the drug.

Assessment of Interindividual Variation. Interindividual variation of a new drug-metabolizing activity can be evaluated in a panel of liver microsomes from patients of different age, sex, drug exposure, or with a particular genotype/phenotype for a genetic polymorphism. A panel of ten livers often permits an estimate of the variation to be expected in vivo.

3.7 Limitations of In Vitro Approaches

The major limitation of all in vitro approaches is that the therapeutic concentration of a new drug and of its primary metabolites or its concentration in a given tissue often is not exactly known. The prediction of the proportion of total clearance that may be influenced by a single enzyme is therefore often based on assumptions, particularly if multiple pathways are involved. The best results of the prediction of in vivo disposition from in vitro data are therefore achieved if a single or only a few enzymes are responsible for a large part of the total clearance in vivo. Under these conditions both polymorphic metabolism and interactions have been reliably predicted (see for instance Gascon and Dayer 1991).

3.8 Examples

We have used the described approach to study the mechanisms of common genetic polymorphisms of drug metabolism (Meyer et al. 1990). Moreover, these methods have been successful in the identification of the enzymes responsible for the metabolism of cyclosporine (Kronbach et al. 1988), midazolam (Kronbach et al. 1989), lidocain (Bargetzi et al. 1989), omeprazole (Anderson et al. 1993), and several other drugs.

3.9 Outlook

The study of the metabolism of new drugs in vitro at an early stage of drug development has repeatedly provided important information regarding potential problems with polymorphisms drug interactions or other forms of interindividual variation long before administration to man or even animals. This allows the planning of more informative early clinical studies and should reduce the time from discovery to market of new drugs. Moreover, in view of the pronounced species differences in drug metabolism, in vitro drug metabolism data may be helpful for selection of the best animal model for toxicity or efficiency studies. As knowledge on the active sites of different cytochromes P450 continues to be generated, it may ultimately be possible to predict substrate selectivity from the chemical structure of new compounds.

References

Andersson T, Miners JO, Veronese ME, Tassaneeyakul W, Meyer UA, Birkett DJ (1993) Identification of human liver cytochrome P450 isoforms mediating omeprazole metabolism. Br J Clin Pharmacol 36:521–530

Bargetzi MJ, Aoyama T, Gonzalez FJ, Meyer UA (1989) Lidocaine metabolism in human liver by cytochrome P450IIIA4 (PCN1). Clin Pharmacol Ther 46:521–527

Broly F, Gaedigk A, Heim M, Eichelbaum M, Mörike K, Meyer UA (1991) Debrisoquine/sparteine hydroxylation genotype and phenotype: analysis of common mutations and alleles of CYP2D6 in a European population. DNA Cell Biol 10:545–558

Daly AK, Cholerton S, Gregory W, Idle JR (1993) Metabolic polymorphisms. Pharmacol Ther 57:129–160

DeMorais SMF, Wilkinson GR, Blaisdell J, Nakamura K, Meyer UA, Goldstein JA (1994) The major genetic defect responsible for thepolymorphism of S-mephenytoin metabolism in humans. J Biol Chem 269:15422–15422

Fonné-Pfister R, Meyer UA (1988) Xenobiotic and endobiotic inhibitors of cytochrome P450db1 function, the target of the debrisoquine/sparteine type polymorphism. Biochem Pharmacol 37:3829–3835

Gascon MP, Dayer P (1991) In vitro forecasting of drugs which may interfere with the biotransformation of midazolam. Eur J Clin Pharmacol 41:573–578

Goldstein JA, Faletto MB, Romkes-Sparks M, Sullivan T, Kitareewan S, Raucy JL, Lasker JM, Ghanayem BI (1994) Evidence that CYPC19 is the major s-mephenytoin 4'hydroxylase in humans. Biochemistry (in press)

Gonzalez FJ (1992) Human cytochromes P450: problems and prospects. Trends Pharmacol Sci 13:346–352

Guengerich FP, Müller-Enoch D, Blair IA (1986) Oxidation of quinidine by human liver cytochrome P-450. Mol Pharmacol 30:287295

Heim M, Meyer UA (1990) Genotyping of poor metabolizers of debrisoquine by allele-specific PCR amplification. Lancet 336:529–532

Johansson I, Lundqvist E, Bertilsson L, Dahl M-L, Sjoeqvist F, Ingelman-Sundberg M (1993) Inherited amplification of an active gene in the cytochrome P450 CYP2D locus as a cause of ultrarapid metabolism of debrisoquine. Proc Natl Acad Sci USA 90:11825–11829

Kalow W (1992) Pharmacogenetics of drug metabolism international encyclopedia of pharmacology and therapeutics (executive editor AC Sartorelli). Pergamon, London

Kalow W, Goedde HW, Agarwal DP (1986) Ethnic differences in reactions to drugs and xenobiotics. Liss, New York

Kronbach T, Fischer V, Meyer UA (1988) Cyclosporine metabolism in human liver: identification of a cytochrome P450 of the P450III gene family as the major cyclosporine-metabolizing enzyme explains interactions of cyclosporine with other drugs. Clin Pharmacol Ther 43:630–635

Kronbach T, Mathys D, Umeno M, Gonzalez FJ, Meyer UA (1989) Oxidation of midazolam and triazolam by human liver cytochrome P450IIIA4. Mol Pharmacol 36:89–96

Meier PJ, Mueller HK, Dick B, Meyer UA (1983) Hepatic monooxygenase activities in subjects with a genetic defect in drug oxidation. Gastroenterology 85:682–692

Meyer UA (1994) Pharmacogenetics: the slow, the rapid, and the ultrarapid. Proc Natl Acad Sci USA 91 (in press)

Meyer UA, Zanger UM, Grant D, Blum M (1990) Genetic polymorphisms of drug metabolism. In: Testa B (ed) Advances in drug research, vol 19. Academic, London, pp 307–23

Nelson DR, Kamataki T, Waxman DJ, Guengerich FP, Estabrook RW, Feyereisen R, Gonzalez FJ, Coon MJ, Gunsalus IC, Gotoh O, Okuda K, Nebert DW (1993) the P450 superfamily: update on new sequences, gene mapping, accession numbers, early trivial names of enzymes, and nomenclature. DNA Cell Biol 12:1–51

Okey AB (1990) Enzyme induction in the cytochrome P450 system. Pharmacol Ther 45:241–298

Waxman DJ, Azaroff L (1992) Phenobarbital induction of cytochrome P450 gene expression. Biochem J 281:577–592

Whitlock JP (1993) Mechanistic aspects of dioxin action. Chem Res Toxicol 6:754–763

Wilkinson GR, Guengerich FP, Branch RA (1992) Genetic polymorphism of S-mephenytoin hydroxylation. In: Kalow W (ed) Pharmacogenetics of drug metabolism, vol 2. Pergamon, New York, 657–685

Zanger UM, Vilbois F, Hardwick J, Meyer UA (1988) Absence of hepatic cytochrome P450buf1 causes genetically deficient debrisoquine oxidation in man. Biochemistry 27:5447–5454

4 Metabolic Activation of Anticancer Oxazaphosphorines by Cytochrome P450s: Development of a Model for Cancer Gene Therapy

L. Chen and D. J. Waxman

4.1   Introduction   57
4.2   Metabolism of Oxazaphosphorines
        by Hepatic Cytochrome P-450s   59
4.3   Stable Transfection of Cytochrome P450 cDNAs into Cultured
        Tumor Cells   64
4.4   Cytochrome P450 as a Model for Cancer Treatment
        by Gene Therapy   70
4.5   Evaluation of the Bystander Killing Effect   73
References   78

4.1 Introduction

Cyclophosphamide and its isomer ifosphamide are cell cycle-nonspecific alkylating agents that have a broad spectrum of activity against human cancers and are widely used in cancer chemotherapy (reviewed by Sladek 1988). Cyclophosphamide and ifosphamide are therapeutically inactive prodrugs that must be activated by liver metabolism following administration to cancer patients. Early studies implicated liver cytochrome P450 as the catalyst of cyclophosphamide and ifosphamide activation, primarily on the basis of the observed decrease in drug elimination half-life after administration of phenobarbital or pred-

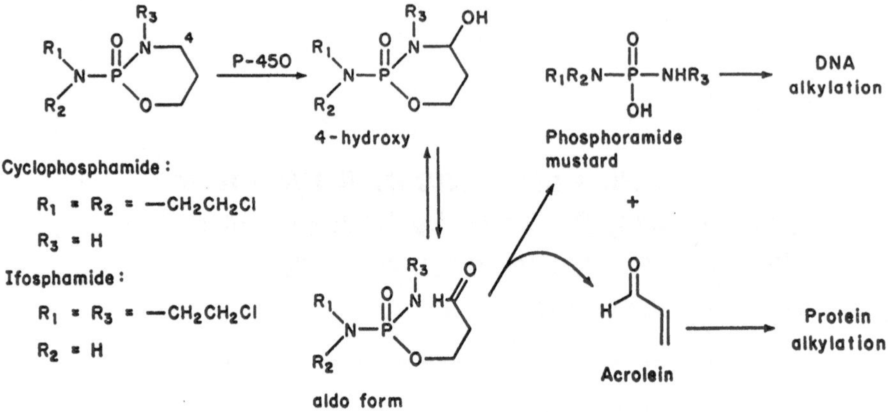

Fig. 1. Pathways of cytochrome P450-catalyzed cyclophosphamide and ifosphamide metabolism

nisone (Faber et al. 1974; Sladek 1972), agents that are known to induce
liver P450 enzymes both in humans and in rats (Pichard et al. 1992).
More recent studies carried out in this laboratory have validated this
conclusion using purified and cDNA-expressed P450 enzymes, as summarized below (Chang et al. 1993; Clarke and Waxman 1989).

Cyclophosphamide and ifosphamide are hydroxylated by cytochrome P450 to yield 4-hydroxycyclophosphamide and 4-hydroxyifos-
phamide, respectively, which exist in equilibrium with the corresponding ring-opened aldophosphamides (Fig. 1). These primary metabolites
then undergo spontaneous β-elimination to yield acrolein and an electrophilic mustard (phosphoramide mustard or ifosphoramide mustard)
in equimolar amounts. The mustards possess DNA-alkylating activity
and are generally considered to be the therapeutically active metabolites
of these oxazaphosphorine anticancer drugs (Sladek 1988). Acrolein, an
electrophilic aldehyde, is highly reactive and can bind covalently to
form protein adducts. The primary 4-hydroxycyclophosphamide/aldo-
phosphamide can also be oxidized to a noncytotoxic metabolite, carbox-
yphosphamide, in a reaction catalyzed by NAD-linked aldehyde dehydrogenase, which can confer drug resistance (Manthey et al. 1990).
Another pathway of oxazaphosphorine metabolism involves N-dechloroethylation of the parent drugs; this pathway yields monofunctional

(non-DNA cross-linking) metabolites that are therapeutically ineffective. The N-dechloroethylation pathway is quantitatively minor for cyclophosphamide, but is major in the case of ifosphamide, where it can consume up to half of the standard drug dose given to cancer patients. The N-dechloroethylation pathway is not only associated with drug inactivation; it also results in the production of chloroacetaldehyde, a neurotoxic metabolite (Norpoth 1976).

Although the metabolism of cyclophosphamide and its isomeric analog ifosphamide have been studied extensively in vivo and in vitro for many years, it was only recently that the specific liver cytochrome P450 enzyme catalysts of drug activation were identified in both rat (Clarke and Waxman 1989; Weber and Waxman 1993) and human liver (Chang et al. 1993). These studies thus provide an underlying basis for the design of rational strategies (a) to enhance or, alternatively, to decrease systemic drug activation through modulation of the activities of individual liver cytochrome P450 enzymes and (b) to sensitize tumor cells to oxazaphosphorines by transfer of prodrug-activating cytochromes P450 into tumor cells. The present contribution summarizes our recent studies using 9L gliosacoma as a model to investigate the feasibility and therapeutic utility of employing cytochrome P450 for cancer gene therapy. The ultimate goal of these studies is to achieve optimal drug efficacy by increasing oxazaphosphorine specificity and selectivity, while minimizing systemic toxicity.

4.2 Metabolism of Oxazaphosphorines by Hepatic Cytochrome P-450s

In our earlier studies we demonstrated that three specific rat liver cytochrome P450 enzymes, CYP2B1 (phenobarbital inducible), CYP2C6 (constitutively expressed), and CYP2C11 (constitutively expressed and present only in adult male rats), are the major catalysts of cyclophosphamide activation in adult rat liver (Clarke and Waxman 1989). [NB: Individual cytochrome P450 enzymes are designated according to the systematic gene nomenclature for CYPs (cytochrome P450s) (Nelson et al. 1993).] As shown in Fig. 2a, about 80% of cyclophosphamide activation catalyzed by phenobarbital-induced adult male rat liver microsomes can be specifically inhibited by an antibody raised against puri-

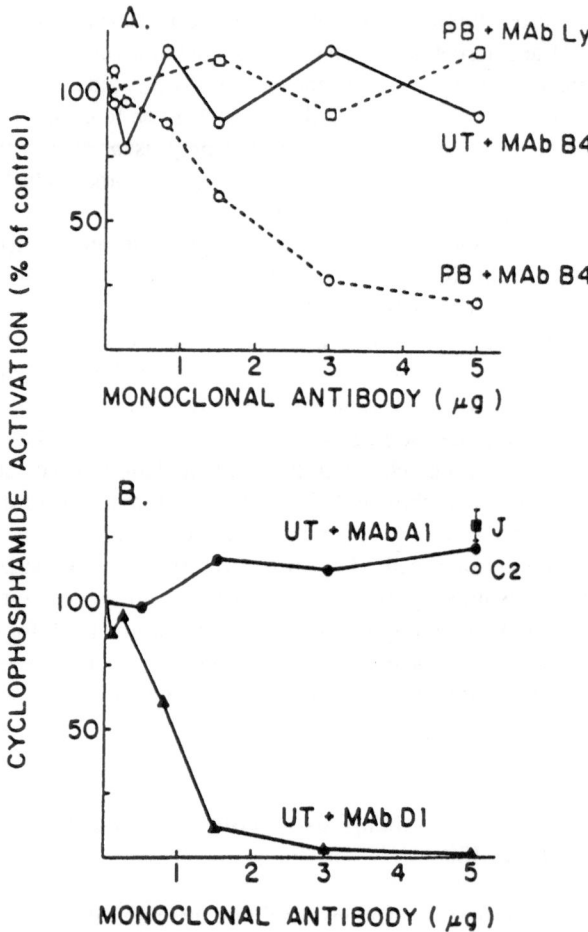

Fig. 2 a, b. Selective inhibition of rat liver microsomal cyclophosphamide activation by P450 form-specific antibodies. Liver microsomes prepared from uninduced (*UT*; *solid lines*) or phenobarbital-induced (*PB*; *dashed line*) adult male rats were incubated with varying amounts of purified monoclonal antibodies reactive with P450 2B1 (*MAb B4*), lysozyme (*MAb Ly*; used as a control for the effects nonspecific antibody; **a**) or P450s 2C6 + 2C11 (*MAb D1*), P450s 1A1 + 1A2 (*MAb A1*), P450 2E1 (*MAb J*) P450 3A (*MAb C2*; **b**). Residual cyclophosphamide activation activity was measured in comparison to samples incubated without antibody. (Data from Clarke and Waxman 1989)

fied CYP2B1, whereas little or no inhibition is obtained using anti-bodies reactive with several other rat hepatic cytochromes P450. How-ever, cyclophosphamide activation catalyzed by uninduced male rat liver microsomes, which do not contain significant amounts of CYP2B1, and which metabolize cyclophosphamide at a much lower rate than phenobarbital-induced microsomes, showed little inhibition by anti-CYP2B1 antibodies, but near complete inhibition (95%) by anti-bodies reactive with both CYP2C6 and CYP2C11 (Fig. 2b). In a sub-sequent study (Weber and Waxman 1993) we demonstrated that the same three rat P450 enzymes also catalyze ifosfamide activation, as judged by P450 induction studies and by antibody and chemical inhibi-tion experiments. In addition, CYP3A enzymes were identified as major catalysts of ifosfamide 4-hydroxylation in rat liver microsomes (Weber and Waxman 1993).

To better understand the role of specific human cytochrome P450 enzymes in oxazaphosphorine activation we extended our studies to human P450s with the goal of identifying the specific P450 enzymes that catalyze cyclophosphamide and ifosfamide 4-hydroxylation in human liver tissue (Chang et al. 1993). Steady-state enzyme kinetic analysis using human liver microsomes revealed that the activation of cyclophosphamide and ifosfamide is catalyzed by both high affinity ($K_m \leq 100\ \mu M$) and low affinity ($K_m \geq 1\ mM$) cytochrome P450 enzymes. As summarized in Table 1, which presents the oxazaphos-phorine activation activities of a panel of microsomes prepared from human B-lymphoblastoid cell lines that were stably transfected with individual cytochrome P450 cDNAs (Crespi 1991), cDNA-expressed human P450 enzymes CYP2A6, 2B8, 2C9 and 3A4 are each catalyti-cally competent with respect to cyclophosphamide and ifosfamide 4-hydroxylation, whereas CYP1A1, 1A2, and 2E1 do not exhibit detect-able activities. Comparison of the activation rates at low and high oxazaphosphorine concentrations suggested that CYP2C8 and 2C9 correspond to low-K_m liver microsomal oxazaphorine 4-hydroxylases, while CYP2A6, 2B6 and 3A4 are high-K_m enzymes.

While these studies establish the inherent catalytic capacities of these individual human P450s with cyclophosphamide and ifosfamide as substrates, they do not reveal which P450 enzymes play a dominant role in oxazaphosphorine metabolism catalyzed by human liver micro-somes, which contain a complex mixture of cytochromes P450. This

Table 1. Oxazaphosphorine activation and xenobiotic metabolism catalyzed by cDNA-expressed human P450 enzymes

	7-Ethoxycoumarin O-deethylation	Cyclophosphamide 4-hydroxylation (pmol product min/mg/protein)		Ifosphamide 4-hydroxylation (pmol product min/mg/protein)	
P450		0.25 mM	2 mM	0.25 mM	2 mM
1A1	36	<1	<1	<1	<1
1A2	17	<1	<1	<1	<1
2A6	50	7	58	3	17
2B6	37	88	764	11	94
2C8	5	16	12	19	4
2C9	7	63	97	31	41
2E1	36	<1	<1	<1	<1
3A4	5	<1	10	<1	18

Microsomes were prepared from human B-lymphoblastoid cells expressing the indicated human P450 cDNAs (Crespi 1991) and then assayed for the oxidation of 7-ethoxycoumarin (1 mM), cyclophosphamide (0.25 mM or 2 mM) or ifosphamide (0.25 mM or 2 mM). Activities shown are corrected for background activities (~ 5 pmol/min/mg protein) measured in microsomes prepared from control (non-P450 expressing) cells. (Data from Chang et al. 1993)

important question was evaluated by the use of P450 form-selective chemical inhibitors as well as P450 subfamily-specific inhibitory polyclonal antibodies. As shown in Table 2, in one human liver microsomal sample, designated HLS9, orphenadrine, a selective CYP2B6 inhibitor, and anti-CYP2B IgG both inhibited cyclophosphamide 4-hydroxylation to a significant extent (about 40% inhibition), while ifosphamide 4-hydroxylation was inhibited less extensively (about 20% inhibition). In contrast, troleandomycin, a CYP3A inhibitor, and anti-CYP3A IgG substantially inhibited ifosphamide 4-hydroxylation (up to 60% inhibition when activity is measured at 2 mM drug substrate) but had little or no effect on cyclophosphamide 4-hydroxylation. By contrast, while

Table 2. Chemical and antibody inhibition of cyclophosphamide and ifosphamide hydroxylation catalyzed by human liver microsomes[a]

	Cyclophosphamide		Ifosphamide	
	0.25 mM	2 mM	0.25 mM	2 mM
HLS9	(0.77)	(4.63)	(0.47)	(3.00)
+Orphenadrine	40	47	21	28
+Anti-CYP2B IgG	38	37	15	20
+Triacetylolendomycin	6	28	23	57
+Anti-CYP3A IgG	12	0	37	63

[a] Results are expressed as percent inhibition of human liver microsomal oxazaphosphorine activation obtained using 0.3 mM orphenadrine or 25 μM triacetyloleandomycin, selective inhibitors of human CYP2B6 and CYP3A enzymes, respectively (Chang et al. 1994). Inhibition data shown for rabbit polyclonal anti-P450 IgGs represent the maximal extent of inhibition of microsomal activity obtained with saturating antibody. Data shown are for human liver microsome sample HLS9, whose uninhibited activity is expressed as nmol product formed/min/mg microsomal protein (values in parenthesis). Data shown are from (Chang et al. 1993)

CYP2A6 can activate these oxazaphosphorine drugs, the protein level of CYP2A6 is apparently too low for it to make a substantial contribution to drug activation in human liver microsomes, as revealed by antibody inhibition studies (Chang et al. 1993). Moreover, the CYP2D6 inhibitor quinidine had little or no inhibitory effect on liver microsomal 4-hydroxylation of either oxazaphosphorine, indicating that CYP2D6 does not contribute to the activation of these drugs in human liver. While the extent to which CYP2B1 and CYP3A enzymes, respectively, contribute to cyclophosphamide and ifosfamide activation is likely to vary from one human liver microsome sample to the next, these and other experiments (Chang et al. 1993) clearly establish the importance of the CYP2B and CYP3A enzymes in the activation of these cancer chemotherapeutic drugs in human liver. Consequently, clinical strategies to improve the therapeutic index of cyclophosphamide and ifosphamide through modulation of liver cytochrome P450 enzyme levels may benefit by focusing on CYP2B and CYP3A as potential targets for modulation of liver metabolic activity through drug induction (Waxman and Azaroff 1992) or other approaches.

4.3 Stable Transfection of Cytochrome P450 cDNAs into Cultured Tumor Cells

Stable transfection of specific cytochrome P450 enzymes into P450-deficient cell lines can be a powerful tool to define the function of individual P450 in the biotransformation of chemotherapeutic agents (LeBlanc and Waxman 1989) or toxins and carcinogens (Doehmer et al. 1988). In our studies we have used human lymphoblastoid cell lines that were stably transfected with individual P450 enzymes by Dr. C. Crespi and associates (Gentest Corporation, Woburn, Massachusetts; Crespi 1991) to study the role of CYP2B6 and CYP2A6 in the activation of cyclophosphamide and ifosphamide (Chang et al. 1993). As shown in Fig. 3, the growth of CYP2B6- and CYP2A6-expressing cells is inhibited by both cyclophosphamide and ifosphamide in a dose-dependent manner. Substantially less growth inhibition was observed in the P450-negative parental cells. These findings are consistent with the requirement of cytochrome P450 to activate these oxazaphosphorines to cytotoxic agents. Co-treatment of the CYP2B6 cells with cyclophosphamide

---

Fig. 3 a–c. Cytotoxicity of cyclophosphamide and ifosphamide toward cultured human B-lymphoblastoid cells expressing CYP2A6 or CYP2B6. Cells stably transformed with cDNA encoding the indicated human P450s or the vector alone (*control*) were treated with the indicated concentrations of cyclophosphamide (*CPA*, **a, b**) or ifosphamide (*IFA*, **c**) as described (Chang et al. 1993). The cytotoxicity of cyclophosphamide is blocked fully by treatment of the CYP2A6-expressing cells (**a**) with coumarin (a CYP2A6 substrate), while 7-ethoxy-4-trifluoro-methylcoumarin (7EFC, a CYP2B6 substrate) partially reduces the cytotoxicity of cyclophosphamide to the CYP2B6-expressing cells (**b**). Results are expressed as the number of cells in cultures treated with cyclophosphamide or ifosphamide relative to the corresponding cell line without drug treatment, and are graphed on a log scale. Control (*open circles*): drug treatment of parental lymphoblastoid cells transformed with vector alone. (Data from Chang et al. 1993)

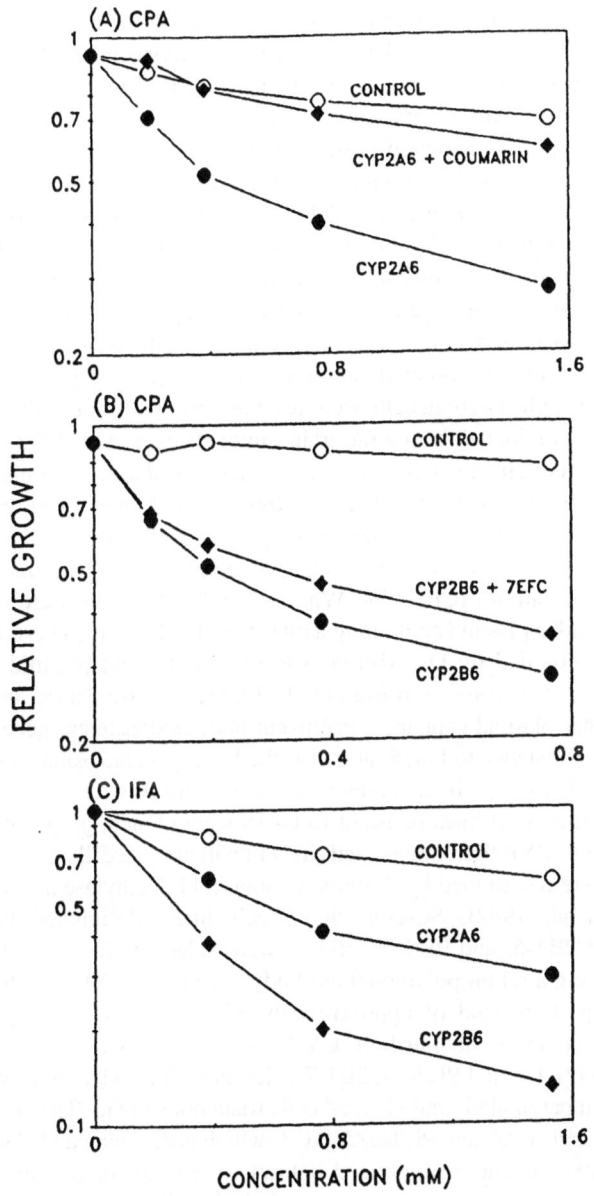

plus the CYP2B6 substrate 7-ethoxy-4-trifluoromethyl-coumarin, or treatment of the CYP2A6 cells with cyclophosphamide plus the CYP2A6 substrate coumarin, resulted in a partial inhibition of cyclophosphamide cytotoxicity (Fig. 3), confirming that the cytotoxicity of cyclophosphamide toward these cells is P450 metabolism dependent. These findings are in agreement with an earlier report that a transgenic *Drosophila* which expresses CYP2B1 is hypersensitive to cyclophosphamide (Jowett and Wajidi 1991). On the basis of these findings we considered the possibility that the transfer of oxazaphosphorine-activating cytochrome P450 genes in tumor cells may render these cells preferentially susceptible to cyclophosphamide and ifosfamide and therefore may have a useful application in cancer chemotherapy.

Rat 9L gliosarcoma cells were used as a model for our initial study. 9L cells, originated from a rat brain tumor (Barker et al. 1973), can be grown in culture or can be implanted either subcutanously or intracranically in Fischer 344 rats. 9L cells express cytochrome P450 reductase but little or no P450 enzymes, making them well suited as a recipient cell line for experiments involving cytochrome P450 gene transfer. In our recent studies (Chen and Waxman 1994), 9L cells were cotransfected with a plasmid containing a rat CYP2B1 cDNA (pMT2-*CYP2B1;* kindly provided by Dr. Milton Adesnik, NYU) and plasmid pCMV β-gal. Neo (a generous gift from Dr. H. Li, Dana Farber Cancer Institute). This latter plasmid contains a neomycin phosphotransferase gene, which confers resistance to G418, and also the lacZ (β-galactosidase) gene of *Escherichia coli*, which serves as a control and provides a convenient cell marker. Cell lines resistant to G418 were cloned, propagated, and evaluated. CYP2B1 enzyme activities in parental 9L cells and transfectants were determined by 7-ethoxycoumarin O-deethylase assay (Waxman et al. 1982). Several clonal cell lines, designated 9L/lacZ, 9L/lacZ/2B1-6, and 9L/lacZ/2B1-7, were selected. Western blot analysis using a rabbit polyclonal antibody specific to CYP2B1 showed a single protein band of approximately 52 kDa, corresponding to the molecular mass of purified CYP2B1, in samples prepared from 9L/lacZ/2B1-6 and 9L/lacZ/2B1-7 cells. No CYP2B1 protein was detected in parental 9L and 9L/lacZ cells (data not shown). The clonal cell lines 9L, 9L/lacZ, and 9L/lacZ/2B1-7 (which is designated 9L/lacZ/2B1 in the experiments summarized below) were used for further studies. P450 enzyme activity assays verified that 9L/lacZ/2B1 expresses

CYP2B1 in an enzymatically active form. P450 2B1 activity was not detectable in parental 9L cells or in 9L/lacZ cells, which express β-galactosidase but not P450 2B1.

We next tested the sensitivity of these cell lines to the cytotoxic effects of cyclophosphamide and ifosphamide. CYP2B1-positive cells (9L/lacZ/2B1) and CYP2B1-negative cells (parental 9L and 9L/lacZ) were cultured with various concentrations of cyclophosphamide or ifosphamide in αMEM containing 10% fetal bovine serum. The number of viable cells present 5 days after drug treatment was then determined. As shown in Fig. 4a, cyclophosphamide inhibited the growth of CYP2B1-positive cells in a concentration-dependent manner. Growth of CYP2B1-positive cells was also inhibited by ifosphamide, but required a somewhat higher drug concentration (Fig. 4b). These finding are consistent with our earlier conclusion that CYP2B1 activates ifosphamide at a somewhat lower rate than cyclophosphamide (see Table 2). In contrast, parental 9L cells and 9L/lacZ cells manifested no adverse effects when grown in the presence of cyclophosphamide or ifosphamide. In control experiments we have shown that CYP2B1-positive and CYP2B1-negative cells are both inherently sensitive to activated cyclophosphamide, which we presented to the cells in the form of 4-hydroperoxy-cyclophosphamide (4HC; Fig. 4c).

To verify that the expression of CYP2B1 per se is responsible for the chemosensitivity of 9L/lacZ/2B1 cells to cyclophosphamide, we used a CYP2B1-specific enzyme inhibitor, metyrapone (Waxman and Walsh 1983), to inhibit the CYP2B1 activity of the CYP2B1-positive cells. In the presence of 10 μM metyrapone, the cytotoxic effect of cyclophosphamide and ifosphamide on 9L/lacZ/2B1 cells was substantially decreased (MTP; Fig. 5). Although metyrapone thus showed a chemoprotective effect on CYP2B1-positive cells treated with cyclophosphamide and ifosphamide, it did not block the cytotoxic effect of the chemically activated derivative 4-hydroperoxy-cyclophosphamide, a finding that is consistent with metyrapone protection via inhibition of P450-catalyzed oxazaphosphorine activation. Therefore the chemosensitivity of 9L/lacZ/2B1 cells to cyclophosphamide and ifosphamide is dependent on the presence of a functional CYP2B1 enzyme in these cells.

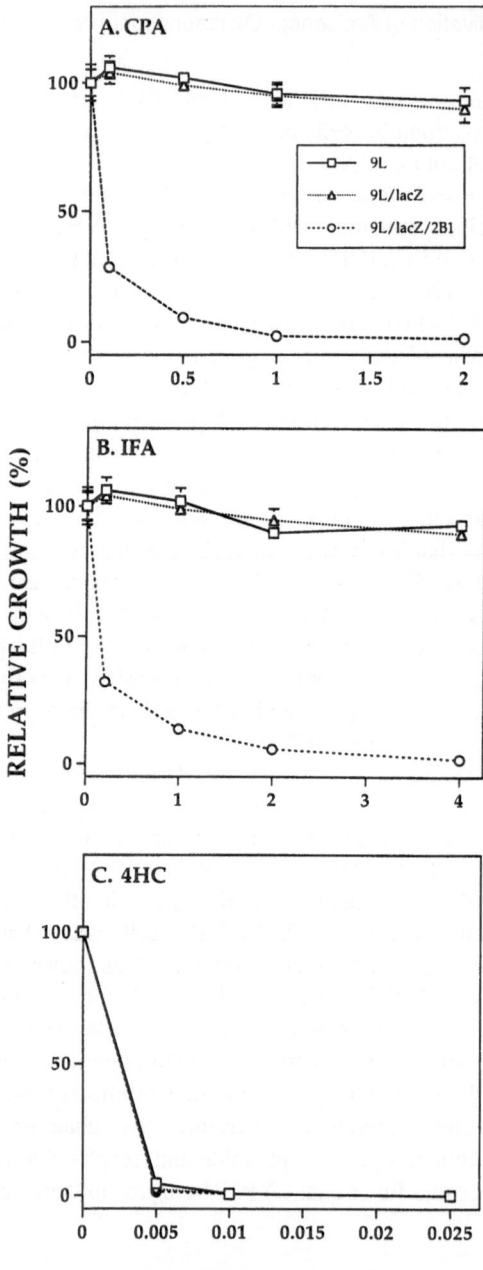

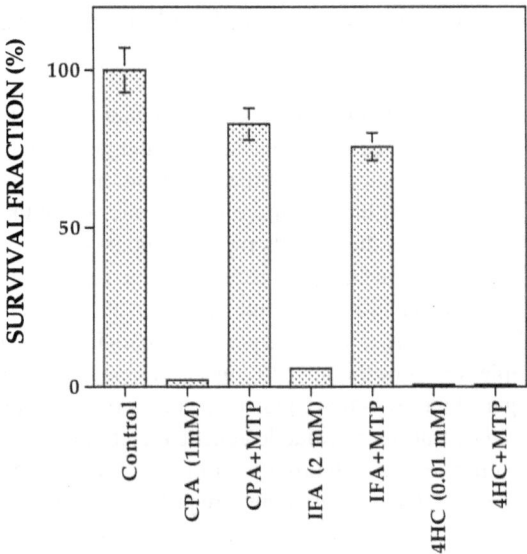

Fig. 5. CYP2B1 enzyme inhibitor metyrapone blocks the cytotoxic effects of cyclophosphamide and ifosphamide but not 4-hydroperoxycyclophosphamide on CYP2B1-positive cells. CYP2B1-positive cells (9L/lacZ/2B1). Cells (1 × 10^5) were plated in duplicate in 30-mm tissue culture plates. Cells were treated for 5 days with either 1 mM cyclophosphamide (*CPA*), 2 mM ifosphamide (*IFA*), or 10 μM 4-hydroperoxycyclophosphamide (*4HC*) in the absence or presence of 10 μM metyrapone (*MTP*), as indicated. Controls received no drug treatment. Cell numbers were determined 5 days after beginning drug treatment. Data represent mean ± range of cells surviving drug treatment compared to cell number in untreated control cultures. (Data from Chen and Waxman 1994)

Fig. 4a–c. Relative survival of CYP2B1-negative cells (parental 9L and 9L/lacZ) and CYP2B1-positive cells (9L/lacZ/2B1) in the presence of cyclophosphamide or ifosphamide. Cells (1 × 10^5) were plated in duplicate in 30-mm tissue culture plates. Cells were treated for 5 days with the indicated concentrations of cyclophosphamide, ifosphamide, or 4-hydroperoxycyclophosphamide (**a–c**, respectively). Surviving cells were counted 5 days after beginning drug treatment. The effect of drugs on cell survival was expressed as a ratio (%) of cell number in plates with drug treatment to that of without drug treatment. (Data from Chen and Waxman 1994)

4.4 Cytochrome P450 as a Model for Cancer Treatment by Gene Therapy

Conventional chemotherapy aims to kill malignant cells without major toxicity to normal host cells and tissues. Although chemotherapy has achieved some notable successes in the treatment of a number of tumor types (e.g., leukemia), it has had more limited success in the treatment of solid tumors due to the low therapeutic index of many cytotoxic drugs as well as the intrinsic or acquired drug resistance that often characterizes tumor cells in the clinic. One novel approach to cancer treatment that has emerged over the past few years involves application of the principles of "gene therapy" to cancer therapeutics. The basic idea underlying gene therapy is that a genetic disorder or an acquired disease may be treated by transfer of suitable genetic material into a specific cell type (either tumor or host) in order to confer a new metabolic or other function on the cell. Diverse strategies are presently under development for the application of gene therapy to cancer treatment:

- Render tumor cells drug sensitive: introduce "suicide" genes for prodrug-activating enzymes, for example:
  - Ganciclovir → [HSV-TK] →phosphorylated forms
  - 5-Fluorocytosine → [cytosine deaminase]→ 5-fluorouracil
  - Cyclophosphamide → [CYP2B1] → phosphoramide mustard
- Restore tumor-suppressor functions: introduce tumor suppressor genes (e.g., p53) or antisense nucleotides that block expression of oncogenes (e.g., K-*ras*).
- Augment tumor immunogenicity and host immune response: introduce interleukin-2, tumor necrosis factor, granulocyte-macrophage colony-stimulating factor into tumor cells or lymphocytes.
- Protect bone marrow cells: introduce multidrug resistance gene (MDR1) to allow higher-dose chemotherapy.

One of these strategies is based on direct transfer of a so-called "suicide gene" which encodes an enzyme that can activate a prodrug within tumor cells, thereby rendering the tumor cells sensitive to chemicals which are otherwise nontoxic to the cell. The best studied example is the herpes simplex virusthymidine kinase (HSV-TK)/ganciclovir system (Borrelli et al. 1988; Culver et al. 1992; Ezzeddine et al. 1991; Moolten and Wells 1990; Ram et al. 1993). Cells transduced with the HSV-TK gene become sensitive to the antiviral drug and nucleoside analog ganciclovir, while normal cells, which do not express the HSV-TK gene, are unaffected (McLaren et al. 1985). Ganciclovir is metabolized by HSV-TK to a product that ultimately yields a phosphorylated nucleotide analog which become incorporated into the DNA strands of cells in the S phase of the cell cycle and subsequently leads to the arrest of DNA synthesis and cell death. The utility of HSV-TK for cancer therapy can be seen by the major tumor regression that results from transduction of HSV-TK into rat 9L gliosarcoma in vivo using a retroviral vector after ganciclovir treatment (Culver et al. 1992). The first gene therapy trial for the treatment of brain tumors using HSV-TK/ganciclovir has recently been approved and is being carried out at the National Institutes of Health.

A second example of a drug-activating gene that may be useful for cancer therapy is the bacterial cytosine deaminase gene/5-fluorocytosine system (Mullen et al. 1992). Cells modified to express cytosine deaminase become sensitive to 5-fluorocytosine due to the formation of 5-fluorouracil in a reaction catalyzed by cytosine deaminase. 5-Fluorouracil is subsequently metabolized to 5-fluoro-2'-deoxyuridine 5'-monophosphate and 5-fluorouridine 5'-triphosphate, which kill cells by a mechanism that involves inhibition of both DNA and RNA synthesis (Damon et al. 1989). Although both the HSV-TK/ganciclovir system as well as the cytosine deaminase/5-fluorocytosine system may be potentially useful for cancer gene therapy, it is important to note that ganciclovir was originally introduced into the clinic for treatment of herpes viral infection (Field et al. 1983; Smee et al. 1980; Smith et al. 1982), while 5-fluorocytosine is an antifungal drug (Bennett 1990). There are therefore few detailed biochemical or pharmacological studies on the application of these drugs in cancer treatment. It is unknown whether the treatment of viral or fungal infection using these drugs may interfere with the efficacy of cancer therapy in patients undergoing HSV-TK or cytosine deaminase gene transfer. Therefore it may be useful to consider other potential "suicide genes" for cancer gene therapy.

The oxazaphosphorines cyclophosphamide and ifosfamide have been used extensively in the clinic for cancer treatment and their efficacy with respect to tumor cell kill following liver P450 metabolism is well established (Sladek 1988). Consequently, cytochrome P450 gene transfer in combination with cyclophosphamide or ifosfamide treatment could correspond to a useful therapeutic strategy to increase the activity and improve the therapeutic index of this class of chemotherapeutic agents. One of the potential advantages of P450 gene transfer in combination with cyclophosphamide and ifosfamide treatment is that this may provide an opportunity to minimize the systemic toxicity associated with conventional oxazaphosphorine treatment, which entails drug activation in the liver followed by release into the circulation of activated drug metabolites. Although these activated metabolites can effect tumor regression, they are accompanied by inevitable cytotoxic side effects toward normal tissues such as bone marrow and kidney. Since tumor cells do not normally express CYP2B1, we anticipate that introduction of the CYP2B gene into tumor cells will sensitize the cells to oxazaphosphorines as a result of the direct activation of the prodrug within the tumor cells. This could provide the opportunity for improved chemotherapeutic efficacy using conventional or perhaps even low doses of oxazaphosphorine.

In our studies (Chen and Waxman 1994) we employed 9L gliosarcoma cells that are stably transfected to express CYP2B1 (see above) as an ex vivo model to evaluate the feasibility of using the CYP2B1/oxazaphophorine system for cancer gene therapy:

1. 9L gliosarcoma – rat brain tumor, readily grown in culture, s.c. and intracranially in F344 rats. Parental 9L tumor cells express P450 reductase but little or no P450.
2. Use neomycin resistance genes to select for genetically modified 9L cells.
3. Assess sensitivity to cyclophosphamide in cell culture.
4. Evaluate potential therapeutic advantages in vitro and in vivo.
5. Examine occurrence of bystander killing effect.
6. Attempt to further augment tumor cell chemosensitivity by overexpression of NADPH cytochrome P450 reductase.

In view of the positive results from our in vitro experiments (Figs. 4, 5) we initiated an in vivo tumor growth delay study to compare the cyclophosphamide sensitivity of parental 9L cells to that of CYP2B1-expressing 9L tumor cells. A preliminary experiment was carried out using parental 9L cells and 9L/2B1-2 cells (a CYP2B1-expressing 9L gliosarcoma-derived cell line kindly provided by Dr. Antonio Chiocca, Massachusetts General Hospital) grown as solid tumors in female Fischer 344 rats. Seven days after tumor implantation cyclophosphamide was administered at a dose of 100 mg/kg body weight by a single i.p. injection. Cyclophosphamide treatment of 9L/2B1-2 tumors led to complete tumor regression (data not shown). By contrast, although 9L tumors also showed some growth delay following cyclophosphamide treatment, this antitumor effect was short term, with aggressive tumor growth eventually returning. The temporary growth delay of 9L tumors is due to the activation of cyclophosphamide by cytochrome P450 present in the liver, which in the case of adult female rats is primarily cytochrome P450 form 2C6 (Clarke and Waxman 1989). Our preliminary results in these in vivo tumor model studies are very exciting, as they indicate that a solid tumor *can* be rendered highly susceptible to oxazaphosphorine treatment in vivo in cases where intratumoral prodrug activation can be achieved by the tumoral expression of the CYP2B1 gene.

4.5 Evaluation of the Bystander Killing Effect

In vitro and in vivo studies of the HSV-TK/ganciclovir system have indicated that not all tumor cells need to be transduced with HSV-TK to achieve high efficiency of tumor cell killing. This finding can be explained by the "bystander effect," whereby HSV-TK transduced cells treated with ganciclovir exert a "bystander killing" of non-HSV-TK transduced cells which they contact (Bi et al. 1993; Culver et al. 1992; Freeman et al. 1993; Ram et al. 1993). This bystander effect can be of great therapeutic significance, because it indicates that eradication of the tumor can, in principle, be achieved even if only a subset of a tumor cell population is effectively transduced with the therapeutic gene. Consequently, we carried out experiments to model whether P450 gene transfer is associated with a bystander effect, i.e., whether P450-expressing cells can sensitize adjacent tumor cells to cyclophosphamide. We first examined whether CYP2B1-negative 9L cells are rendered susceptible to cyclophosphamide cytotoxicity when cocultured with CYP2B1-expressing tumor cells. Equal numbers of CYP2B1-negative (parental 9L) cells were mixed with CYP2B1-positive cells (9L/lacZ/2B1) in culture and then were treated with cyclophosphamide. If cyclophosphamide cytotoxicity were restricted to the CYP2B1-positive cells, one should observe killing of approximately 50% of the cells in the total cell population. On the other hand, if the CYP2B1-positive cells chemosensitize the adjacent CYP2B1-negative cells, both CYP2B1-positive and CYP2B1-negative cells should be killed following treatment of the coculture with cyclophosphamide. Indeed, we found that nearly 80% of the total cell population was eradicated when the mixed culture was exposed to cyclophosphamide (Chen and Waxman 1994). Moreover, a CYP2B1 enzyme inhibitor, metyrapone, could largely abrogate this effect. In contrast, there was no killing of either cell population when 9L/lacZ cells were mixed with parental 9L cells. The cells in the mixed culture showed a similar pattern of sensitivity to ifosphamide, albeit at a somewhat higher drug concentration. These studies demonstrate that CYP2B1-positive cells do confer a bystander killing of CYP2B1-negative cells by a mechanism that involves CYP2B1 enzyme activity.

In other experiments we evaluated the relative survival of CYP2B1-negative and CYP2B1-positive cells when present in a mixed cell population. Cells marked with the *lacZ* gene (β-galactosidase), which can be identified as blue cells after staining the cultures with the β-galactosidase substrate x-gal, were employed to distinguish the two types of cells in culture. Equal numbers of unmodified parental 9L cells were mixed with *lacZ*-marked CYP2B1-positive cells (9L/lacZ/2B1). Five days after cyclophosphamide treatment cells were fixed and stained with x-gal to reveal the blue cells. As illustrated in Fig. 6, only a few CYP2B1-positive cells (blue staining) survived cyclophosphamide treatment. Some CYP2B1-negative cells remained; however, these cells showed severe cytopathic effects. Although the CYP2B1 inhibitor metyrapone protected both cell types from cyclophosphamide killing, microscopic evaluation revealed morphologic distortions in some of the

---

Fig. 6a–c. Histochemical analysis of lac Z-marked CYP2B1-negative and CYP2B1-positive cells in a mixed cell population. Equal numbers of parental 9L cells were mixed with 9L/lacZ/2B1 cells (*lacZ*-marked CYP2B1-positive cells). Cells (1×10^5) were plated in 30-mm tissue culture plates. Cells were treated with 1 mM cyclophosphamide (*CPA*; **b**), 1 mM cyclophosphamide plus 10 μM metyrapone (*MTP*; **c**), or received no drug treatment as a control (**a**). Five days after treatment cells were fixed in 0.5% glutaraldehyde, then stained with x-gal for 4–8 h to visualize the blue cells. (Data from Chen and Waxman 1994)

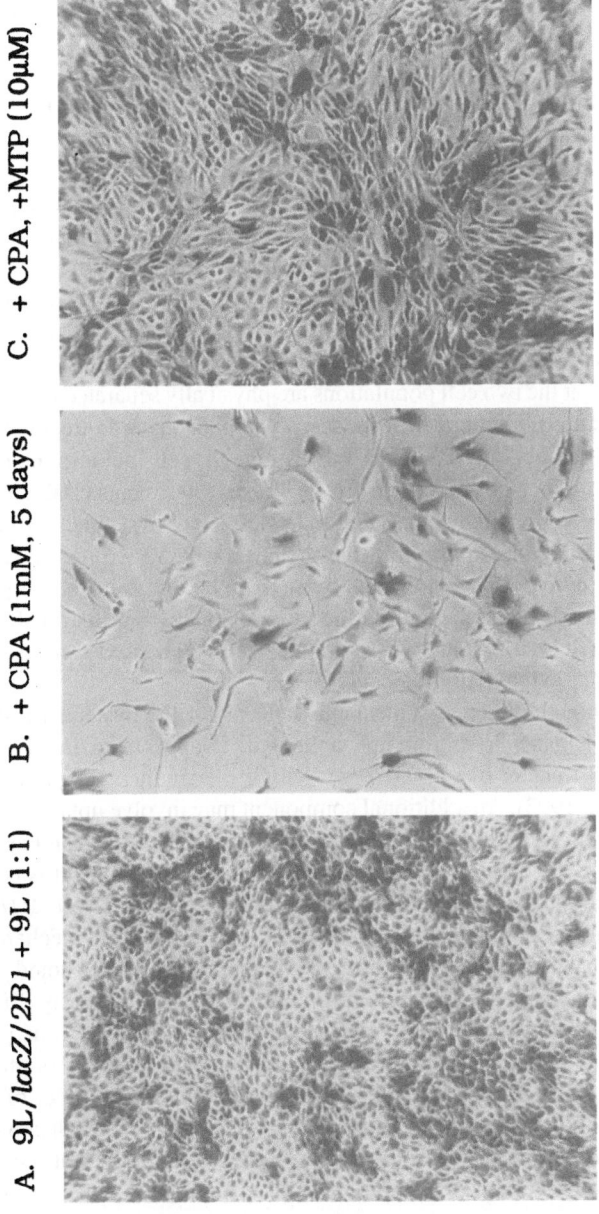

C. + CPA, +MTP (10μM)

B. + CPA (1mM, 5 days)

A. 9L/*lacZ*/2B1 + 9L (1:1)

CYP2B1-positive cells, but not in the CYP2B1-negative cells following treatment with cyclophosphamide in the presence of metyrapone. These findings demonstrate that CYP2B1-positive cells are more susceptible to cyclophosphamide cytotoxicity as a consequence of the prodrug activation that occurs within the tumor cell, but that substantial cytotoxicity toward adjacent CYP2B1-negative cells also occurs.

We also assessed whether the killing of CYP2B1-negative cells mediated by CYP2B1-positive cells requires direct cell-cell contact, by analogy to the case of HSV-TK-positive and HSV-TK-negative tumor cells and ganciclovir treatment (Bi et al. 1993; Culver et al. 1992; Freeman et al. 1993; Ram et al. 1993). For these experiments, parental 9L cells were cocultured with CYP2B1-positive cells (9L/lacZ/2B1) in a way that the two cell populations are physically separated but share the same culture medium. In this case cyclophosphamide treatment for 5 days killed not only the CYP2B1-positive 9L cells but also the parental 9L cells (Chen and Waxman 1994). These experiments establish that in the case of cyclophosphamide and P450 2B1, bystander killing is at least partly due to the transfer to the non-P450 expressing cells of soluble cytotoxic metabolite(s) formed via P450-catalyzed drug activation. This bystander effect is therefore distinct from that of the HSV-TK/ganciclovir system, where intimate cell-cell contact is necessary for bystander cytotoxicity to occur.

Although the precise mechanistic basis for the bystander killing by HSV-TK/ganciclovir remains unknown, it appears to involve direct cell-cell transfer of activated ganciclovir metabolites via gap junctions (Bi et al. 1993). An additional component may involve uptake of phagocytocized vesicles containing activated toxic metabolites and apoptotic signals released by the dying cells (Freeman et al. 1993). It is conceivable that the bystander killing that we observed in the P450 2B1/oxazaphosphorine system could also involve a cell-cell contact mechanism, in addition to the soluble toxic metabolite mechanism noted above. 4-Hydroxycyclophosphamide formed by CYP2B1 is readily diffusible across cell membranes, and it is likely that the release of this primary metabolite, or perhaps its cytotoxic decomposition products phosphoramide mustard and acrolein, contributes to the lethal effect of cyclophosphamide on neighboring CYP2B1-negative cells. Other mechanisms such as the transfer of apoptotic signals from dying cells could also play a role. Further investigations will be required to elucidate in detail the mechanisms of cell death in this ex vivo model of P450 gene transfer.

The lack of a requirement for cell contact to achieve CYP2B1/cyclo-phosphamide bystander cytotoxicity may represent a therapeutic advantage of cytochrome P450 gene therapy over the HSV-TK/ganciclovir system by providing for more extensive distribution of activated drug within a tumor mass owing to the diffusibilty of activated drug metabolites. In addition, unlike HSV-TK/ganciclovir, whose activated metabolites are only cytotoxic to cells in the DNA synthesis (S phase) of the cell cycle, the CYP2B1/oxazaphosphorine system generates metabolites that are effective in a cell cycle-independent manner, resulting in a higher fractional tumor cell kill; the toxicity of phosphoramide mustard-derived interstrand DNA cross-links to tumor cells becomes manifest at whichever point the tumor cells begin to replicate. Another potential advantage of cytochrome P450-based cancer gene therapy is the possibility of augmenting a drug's local anti-tumor effect, via P450 gene transfer, in combination with the selective inhibition of *liver* cytochrome P450s enzymes involved in prodrug activation. This could be achieved by using appropriate P450 form-selective inhibitors to minimize the host tissue toxicity that results from the systemic exposure to activated metabolites which invariably occurs during conventional chemotherapy:

The potential advantages of cytochrome p450-based cancer gene therapy are the following:

– Augmentation of local antitumor effect while minimizing host toxicity due to systemic exposure to activated metabolites.
– Unlike HSV-TK/ganciclovir, P450/cyclophosphamide is active in a cell cycle-independent manner, resulting in a higher fractional cell kill.
– The diffusibility of activated metabolites may sensitize adjacent tumor cells, even in cases where P450 gene transfer transduces only a subset of tumor cell population.
– It is potentially extendable to other chemotherapeutic drugs and P450 genes.

In conclusion, the studies summarized above establish a model system for further study of the therapeutic utility of transferring oxazaphosphorine-activating cytochrome P450 genes into tumor cells. These studies provide evidence that tumor cell kill may proceed in an efficient manner even if only a subset of a tumor cell population is efficiently transfected with the cytochrome P450 gene. A substantial enhancement of cyclophosphamide or ifosphamide cytotoxicity may thus be obtained when using the CYP2B1 gene for cancer therapy, even in the cases where the efficiency of gene transfer *via* viral vectors or other gene transfer approaches is less than 100% with respect to transduction of the P450 gene into the tumor cells. The chemo-gene therapy concepts and strategies developed in this chapter may potentially be extended to other cancer chemotherapeutic agents (LeBlanc and Waxman 1989) and other cytochrome P450 genes (Nelson et al. 1993). Further studies are planned to apply the CYP2B1/oxazaphosphorine system to cancer therapy by transfer of CYP2B1 gene into brain tumors as well as systemic malignant tumors employing a variety of viral vectors.

References

Barker M, Hoshino T, Gurcay O, Wilson CB, Nielsen SL, Downie R, Eliason J (1973) Development of an animal brain tumor model and its response to therapy with 1,3-bis(2-chloroethyl)-1-nitrosourea. Cancer Res 33:976–986

Bennett JE (1990) Goodman and Gilman's the pharmacological basis of therapeutics. Pergamon, New York

Bi WL, Parysek LM, Warnick R, Stambrook PJ (1993) In vitro evidence that metabolic cooperation is responsible for the bystander effect observed with HSV tk retroviral gene therapy. Human Gene Ther 4:725–731

Borrelli E, Heyman R, Hsi M, Evans RM (1988) Targeting of an inducible toxic phenotype in animal cells. Proc Natl Acad Sci U S A 85:7572–7576

Chang TKH, Weber GF, Crespi CL, Waxman DJ (1993) Differential activation of cyclophosphamide and ifosphamide by cytochromes P450 2B and 3A in human liver microsomes. Cancer Res 53:5629–5637

Chang TKH, Gonzalez FJ, Waxman DJ (1994) Evaluation of triacetyloleandomycin, alpha-naphthoflavone and diethyldithiocarbamate as selective chemical probes for inhibition of human cytochromes P450. Arch Biochem Biophys 311:437–442

Chen L, Waxman DJ (1994) Manuscript in preparation

Clarke L, Waxman DJ (1989) Oxidative metabolism of cyclophosphamide: identification of the hepatic monooxygenase catalysts of drug activation. Cancer Res 49:2344–2350

Crespi CL (1991) Expression of cytochrome P450 cDNAs in human B lymphoblastoid cells: applications to toxicology and metabolite analysis. Methods Enzymol 206:123–129

Culver KW, Ram Z, Wallbridge S, Ishii H, Oldfield EH, Blaese RM (1992) In vivo gene transfer with retroviral vector-producer cells for treatment of experimental brain tumors (see comments). Science 256:1550–1552

Damon LE, Cadwin E, Benz C (1989) Enhancement of 5-fluorouracil antitumor effects by the prior administration of methotrexate. Pharmacol Ther 43:155–189

Doehmer J, Dogra S, Friedberg T, Monier S, Adesnik M, Glatt H, Oesch F (1988) Stable expression of rat cytochrome P450IIB1 cDNA in chinese hamster cells (V79) and the metabolic activation of aflatoxin B1. Proc Natl Acad Sci USA 85:5769–73

Ezzeddine ZD, Martuza R L, Platika D, Short MP, Malick A, Choi B, Breakefield XO (1991) Selective killing of glioma cells in culture and in vivo by retrovirus transfer of the herpes simplex virus thymidine kinase gene. New Biol 3:608–614

Faber OK, Mouridsen HT, Skovsted L (1974) The biotransformation of cyclophosphamide in man: influence of prednisone. Acta Pharmacol Toxicol 35:195–200

Field A, Davies M, Dewitt C (1983) 9-[2-Hydroxy-1-(hydroxymethyl)ethoxy]methylguanine: a selective inhibitor of herpes group virus replication. Proc Natl Acad Sci USA 80:4139–4143

Freeman SM, Abboud CN, Whartenby KA, Packman CH, Koeplin DS, Moolten FL, Abraham GN (1993) The "bystander effect": tumor regression when a fraction of the tumor mass is genetically modified. Cancer Res 53:5274–5283

Jowett T, Wajidi MF (1991) Mammalian genes expressed in Drosophila: a transgenic model for the study of mechanisms of chemical mutagenesis and metabolism. EMBO J 10:1075–81

LeBlanc GA, Waxman DJ (1989) Interaction of anticancer drugs with hepatic monooxygenase enzymes. Drug Metab Rev 20:395–439

Manthey CL, Landkamer GJ, Sladek NE (1990) Identification of the mouse aldehyde dehydrogenases important in aldophosphamide detoxification. CR 50:4991–5002

McLaren C, Chen MS, Barbhaiya RH, Buroker RA, Olsen FB (1985) Herpes virus and virus chemotherapy. In: Kono R, Nakajima A (eds) Herpes virus and virus therapy: pharmacological and clinical approaches. Elsevier, Amsterdam, pp 57–61

Moolten FL, Wells JM (1990) Curability of tumors bearing herpes thymidine kinase genes transferred by retroviral vectors. J Natl Cancer Inst 82:297–300

Mullen CA, Kilstrup M, Blaese RM (1992) Transfer of the bacterial gene for cytosine deaminase to mammalian cells confers lethal sensitivity to 5-fluorocytosine: a negative selection system. Proc Natl Acad Sci USA 89:33–37

Nelson DR, Kamataki T, Waxman DJ, Guengerich FP, Estabrook RW, Feyereisen R, Gonzalez FJ, Coon MJ, Gunsalus IC, Gotoh O, Okuda K, Nebert DW (1993) The P450 superfamily: update on new sequences, gene mapping, accession numbers, early trivial names of enzymes, and nomenclature. DNA Cell Biol 12:1–51

Norpoth K (1976) Studies on the metabolism of isophosphamide (NSC-109724) in man. CTR 60:437–443

Pichard L, Fabre I, Daujat M, Domergue J, Joyeux H, Maurel P (1992) Effect of corticosteroids on the expression of cytochromes P450 and on cyclosporin A oxidase activity in primary cultures of human hepatocytes. Mol Pharmacol 41:1047–1055

Ram Z, Culver KW, Walbridge S, Blaese RM, Oldfield EH (1993) In situ retroviral-mediated gene transfer for the treatment of brain tumors in rats. Cancer Res 53:83–88

Sladek NE (1972) Therapeutic efficacy of cyclophosphamide as a function of its metabolism. Cancer Res 32:535–542

Sladek NE (1988) Metabolism of oxazaphosphorines. Pharmacol Ther 37:301–355

Smee D, Martin J, Verheyden J (1980) Anti-herpes virus activity of the acyclic nucleoside 9-(1,3-dihydroxy-2-proposymethyl)guanine. Antimicrob Agents Chemother 23:676–682

Smith KO, Galloway KS, Kennell WL (1982) A new nucleoside analog, 9-[2-hydroxy-1-(hydroxymethyl)ethoxy]methylguanine, highly active in vitro against herpes simplex virus types 1 and 2. Antimicrob Agents Chemother 22:55–61

Waxman DJ, Azaroff L (1992) Phenobarbital induction of cytochrome P-450 gene expression. Biochem J 281:577–592

Waxman DJ, Walsh C (1983) Cytochrome P-450 isozyme 1 from phenobarbital-induced rat liver: purification, characterization, and interactions with metyrapone and cytochrome b5. Biochemistry, 22:4846–4855

Waxman DJ, Light DR, Walsh C (1982) Chiral sulfoxidations catalyzed by rat liver cytochromes P-450. Biochemistry 21:2499–2507

Weber GF, Waxman DJ (1993) Activation of the anti-cancer drug ifosphamide by rat liver microsomal P450 enzymes. Biochem Pharmacol 45:1685–1694

5 The Use of Bacteria for Cytochrome P450 Expression

M. R. Waterman

5.1   Introduction   81
5.2   Development of Bacterial Systems for Overexpression
       of P450s   82
5.3   Variables for P450 Expression in *E. coli*   83
5.4   Use of Bacterial Expression to Study Enzymatic Activities
       of P450s   86
5.5   Use of Bacterial Expression to Study P450 Structure/Function . . .   90
5.6   Disadvantages of Bacterial Expression of P450s   92
5.7   Conclusions   93
References   94

5.1 Introduction

The important research problems in the area of P450 to be addressed during the last decade of the twentieth century include: (a) identification and localization of the approximately 200 forms of human P450, (b) characterization of the enzymatic properties of these enzymes and thus their relatedness regarding overlapping substrate metabolism, and (c) determination of the structure-function relationships in human P450s. Over the past decade revolutionary developments have taken place in both molecular biology and P450 research which have allowed these general questions to be addressed. In particular, the application of heterologous expression systems to P450 research and the advent of polymerase chain reaction technologies have been a key. A number of different expression systems including yeast (Oeda et al. 1985), COS cells (Zuber et al. 1986), baculovirus (Asseffa et al. 1989), vaccinia virus (Gonzalez et al. 1991), permanent cell lines (Doehmer and Oesch 1991; Crespi 1991), and *E. coli* (Barnes et al. 1991) have all proved useful for elucidating the activities of specific forms of P450, both from humans and from other species (Waterman 1994). Many important experiments investigating P450 activities using purified preparations have been carried out since the pioneering P450 purification and reconstitution studies of Lu and Coon (Lu et al. 1969). However, purification is often tedious and sometimes unsuccessful because of the low levels of expression of many forms of P450, and, furthermore, human tissues have frequently not been available. Thus the development of systems which can be used for expression of recombinant enzymes has opened the door to investigation of enzymatic properties of forms of P450 which were previously not available for study. Each of the different heterologous expression systems has its own specific advantages and disadvantages. The purpose of the present contribution is to elaborate those of the *E. coli* system.

5.2 Development of Bacterial Systems for Overexpression of P450s

The initial reports of successful expression of functional forms of P450 in *E. coli* by Larson et al. (1991) and Li and Chiang (1991) did not describe high level expression, for example, the ability to determine P450 concentration by spectrophotometric analysis in intact bacterial cultures. Our laboratory succeeded in such high level expression of both microsomal (Barnes et al. 1991) and mitochondrial (Wada et al. 1991) forms of P450 in *E. coli*. For reasons which we do not yet understand it was necessary to utilize different vectors for overexpression of the two classes (microsomal and mitochondrial) of P450.

As with most mitochondrial proteins, mitochondrial P450s including cholesterol side-chain cleavage cytochrome P450 (P450scc) are encoded by nuclear genes as higher molecular weight precursor proteins whose amino terminal presequences are removed proteolytically upon uptake into mitochondria (DuBois et al. 1981). For expression of P450scc in *E. coli* we removed the coding region for this presequence of bovine P450scc, changed the codon for amino acid 2 from Ile to Val to create a restriction site at the amino terminus, and inserted the cDNA encoding this mature form of P450scc into pTrc99A. (This vector is described in detail on page 139 of the Pharmacia Biotech Molecular and Cell Biology Catalog 1994.) For microsomal P450s we found the pCWori$^+$ vector (Fig. 1b) to be effective. Our initial efforts at expression of microsomal P450s focused on use of the cDNA encoding bovine 17α-hydroxylase cytochrome P450 (P450c17). Use of the native coding sequence for this P450 was not successful, no P450c17 being detectable by either immunoblot or spectrophotometric analysis. However, by making specific modifications in the amino terminal coding sequence high levels of expression of P450c17 could be obtained (Barnes et al. 1991). The modifications leading to overexpression are seen in Fig. 2 and include a single amino acid change at position 2 (Trp→Ala). Studies on requirements for heterologous expression of proteins in *E. coli* have shown that the second codon is very important. Many second codons, including the native TGG (Trp) of bovine P450c17, are very poor for expression in *E. coli* while a few, including GCT (Ala), are very good. Also, reduction of the tendency for the RNA product to form secondary structure by increasing the AT richness of the 5'-end of the sequence enhances the potential for translation, presumably by stabilizing initiation. Thus several silent changes (Fig. 2) were also inserted at the 5'-end of the resultant RNA thereby optimizing the opportunity of translation.

Both plasmids, upon transformation into *E. coli*, led to spectrally detectable levels of P450. Optimal P450scc levels obtained have been about 250 nmol/l bacterial culture while levels of P450c17 above 500 nmol/l culture have been obtained.

5.3 Variables of P450 Expression in *E. coli*

High level expression of functional forms of both mitochondrial and microsomal P450s was achieved by lowering the temperature of the bacterial growth below the normal 37°C and by inducing P450 expression with IPTG for 24–48 h. Temperatures in the range of 28–32°C were optimal for production of functional enzymes. In this temperature range, the P450 was found exclusively in the membrane fraction of *E.*

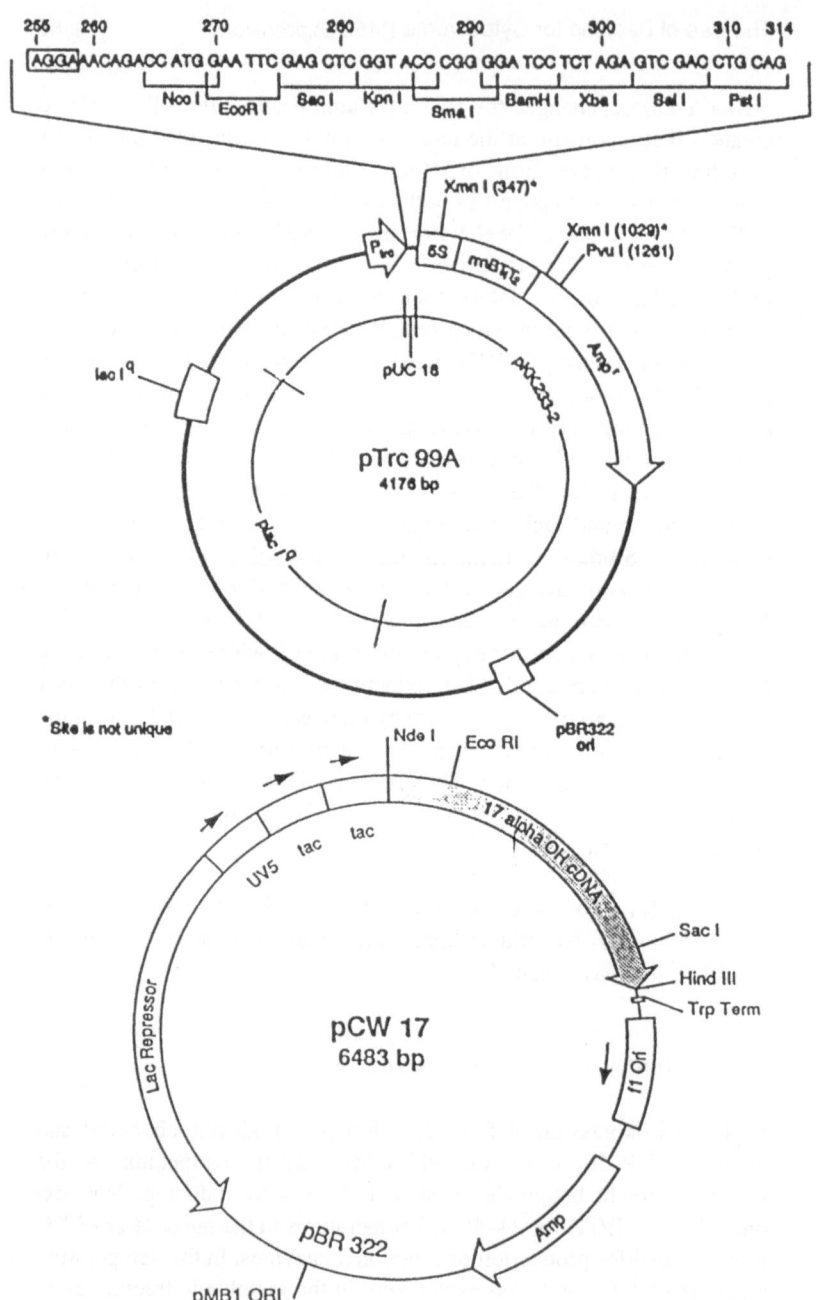

```
      255   260          270          280          290          300          310   314
      ┌─────┐
      │AGGA│AACAGACC ATG GAA TTC GAG CTC GGT ACC CGG GGA TCC TCT AGA GTC GAC CTG CAG
      └─────┘
             Nco I        Sac I   Kpn I          BamH I  Xba I   Sal I    Pst I
                  EcoR I                  Sma I
```

Xmn I (347)*

Xmn I (1029)*
Pvu I (1261)

P_trc   5S   rrnBT₁T₂   Amp^r

lac I^q

pUC 18

pTrc 99A
4176 bp

pKK233-2

plac/q

pBR322 ori

*Site is not unique

Nde I   Eco RI

UV5   tac   tac   17 alpha OH cDNA

Lac Repressor

Sac I
Hind III
Trp Term

pCW 17
6483 bp

f1 Ori

Amp

pBR 322

pMB1 ORI

Bovine nat17	ATG	TGG	CTG	CTC	CTG	GCT	GTC	TTT
	Met	Trp	Leu	Leu	Leu	Ala	Val	Phe
Modified I	ATG	GCT	CTG	TTA	TTA	GCA	GTT	TTT
	Met	Ala	Leu	Leu	Leu	Ala	Val	Phe

Fig. 2. Changes made in the amino-terminal coding sequence (*Modified 1*) leading to overexpression of P450c17 in *E. coli.* The bases which are underlined are those which were changed. Note that the change in codon 2 leads to the amino acid change Trp Ala at position 2

coli. Higher temperatures led to a significant amount of the protein being found in a nonfunctional form in bacterial inclusion bodies. The functional P450s in the *E. coli* membranes (both mitochondrial and microsomal forms) are integral membrane proteins as determined by sodium carbonate treatment, a result also found for the naturally occurring enzymes in either the endoplasmic reticulum or the inner mitochondrial membrane.

An important variable for the expression of P450s in *E. coli* is the bacterial strain. Certain strains, in particular JM109, are very good for expression while others are very poor. For example, JM109 and DH5 produce high levels of functional P450s while DH1, the parent strain of both DH5 and JM109, is very poor for P450 synthesis. The strain XL-1 blue is somewhere in between in its capacity to produce enzymatically active P450s.

We have not exhaustively studied all the parameters which can effect mammalian P450 biosynthesis in *E. coli.* Our impression is that vectors with quite weak promoters (particularly those seen in Fig. 1), optimization of the amino terminal coding region of the P450 cDNA, reduced temperature and specific bacterial strains are all important considerations for achieving high level expression of functional forms of P450 in bacteria.

Fig. 1. pcWori+ – the vector used for expression of the microsomal P450, bovine P450c17. This vector has subsequently been used for the successful overexpression of several functional microsomal P450s in *E. coli*

5.4 Use of Bacterial Expression To Study Enzymatic Activities of P450s

One of the important reasons for the development of heterologous systems for the study of P450s has been the need to investigate the enzymatic properties of single forms of P450 in the absence of other forms of P450. *E. coli* are particularly good for this purpose because of the absence of detectable levels of endogenous P450s in this bacterium. However, bacteria have the limitation that they do not contain the endogenous reductase systems found in animal cells which support P450 activities; NADPH cytochrome P450 reductase in the endoplasmic reticulum and ferrodoxin and ferridoxin reductase in mitochondria.

Mitochondrial P450s:
NADPH → Ferridoxin (adrenodoxin) reductase → Ferrodoxin (adrenodoxin) → P450
Microsomal P450s:
NADPH → NADPH cytochrome P450 REDUCTASE → P450

E. coli contain no reductase system which supports the activity of mitochondrial P450s (Wada et al. 1991). The only way to investigate the activity of these enzymes is to reconstitute their activities in partially or fully purified P450 samples. For example, the membrane fraction from *E. coli* expressing bovine P450scc has been solubilized with detergent and used with purified bovine adrenodoxin and adrenodoxin reductase to achieve high levels of cholesterol side-chain cleavage activity (Table 1). Also this enzyme has been purified to homogeneity from *E. coli* by procedures similar to those used for purification of P450scc from mitochondria and the activity reconstituted in a standard way by addition of adrenodoxin and adrenodoxin reductase (Wada and Waterman 1992). These approaches work for evaluation of enzymatic properties of any form of mitochondrial P450 expressed in a bacteria; however, to date P450scc is the sole mitochondrial P450 expressed in this heterologous system.

In the case of expression of microsomal forms of cytochrome P450 in *E. coli*, quite a different picture develops. First, much to our surprise, microsomal P450s expressed in *E. coli* are active in the intact bacterial cells (Barnes et al. 1991). This was unexpected because it had previously been reported that no protein related to NADPH cytochrome

Table 1. Recombinant P450scc activity[a] (percentage) in *E. coli* membranes

	Expt. 1	Expt. 2
Nontransformed *E. coli*	0.7	0.4
Transformed *E. coli*	0.9	0.6
Complete system-NADPH	n.d[b]	0.6
Complete system[c]	5.9	8.3
Complete system + 0.1% emulgen 913	64.7	52.2

[a] % conversion of 5 mM 25-hydroxycholesterol to prenenolone
[b] not determined
[c] complete system - *E. coli* containing 0.1 nmol bovine P450scc/4.8 g protein in a 1 ml reaction volume containing 5 M of 25-[6-^{3}H]hydroxycholesterol, 0.3% Tween 20, 10 M of adrenodoxin, 1 M NADPH-adrenodoxin reductase, 5 mM glucose 6-phosphate, 0.5 units of glucose-6-phosphate dehydrogenase, and 100 M of NADPH incubated at 37°C for 10 min

P450 reductase could be detected by immunoblot analysis in *E. coli* (Porter et al. 1987). However, pregnenolone and progesterone which are steroid substrates for P450c17 were converted to the expected products when added to intact bacteria expressing bovine P450c17. The turnover numbers in intact *E. coli* are approximately 2% of the values achieved in reconstitution systems, but all expected activities of P450c17 are readily measurable (Fig. 3). Isolation of bacterial membranes containing heterologously expressed P450s demonstrate expected activities when supplemented with purified NADPH cytochrome P450 reductase. They show no activity, however, in the absence of added reductase. Fractionation of *E. coli* indicates that the endogenous reductase which supports microsomal P450 activities is localized in the cytosol and is heat labile. As an aside, the fact that this reductase system is in the cytosol and P450 activities are detectable in intact bacteria established that the expressed P450 is localized on the internal side of the inner bacterial membrane (facing the cytosol). Thus recombinant microsomal P450s are active at a modest rate in intact bacteria, achieve activities equal or greater than those found in microsomes when *E. coli* membranes are reconstituted with excess purified P450 reductase, and can be purified to homogene-

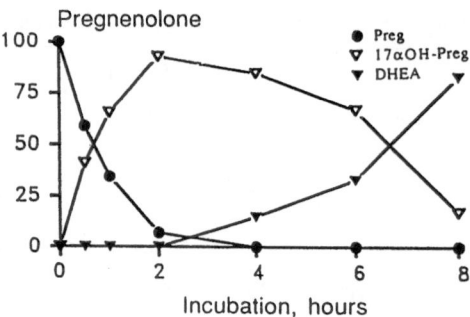

Fig. 3. The activities of recombinant bovine P450c17 expressed in *E. coli.*
2.5μm pregnenolone was added directly to the bacterial culture, and activities
were measured by HPLC as described in Barnes et al. (1991). The conversion of
pregnenolone to 17α-hydroxypregnenolone which in turn is converted to dehydroepiandrosterone is the activity expected for bovine P450c17

ity from *E. coli* and found to be indistinguishable from naturally occurring, purified cytochromes P450.

One reason for expressing cytochromes P450 in bacteria is the aim
of achieving bioreactors containing "designer P450s"; forms of P450
whose catalytic activities have been modified by site-directed mutagenesis in order to engineer catalysis of a particular reaction of commercial
interest. Thus it is necessary to produce intact organisms containing
functional P450s. Relying on the endogenous activity noted above is
inadequate for this purpose because of the low turnover numbers. One
approach pioneered in *E. coli* by Estabrook and colleagues is to construct chimeric proteins between P450s and P450 reductase (Fisher et al.
1992). This approach stems from the discovery that a soluble bacterial
P450, P450BM3, is such a fusion (Narhi and Fulco 1987). The turnover
numbers for P450 BM3 activities are 10–1000 times greater than found
for reconstituted eukaryotic forms of P450. This theme is repeated in
mammalian systems in the enzyme, nitric oxide synthase. Using recombinant techniques, Okahwa and colleagues produced such fusion forms
of P450 and P450 reductase which were fractional and produced turnover numbers ten times greater than the P450 molecule alone when
expressed in yeast (Yabusaki et al. 1988). Thus expression of such
fusion proteins is one way of increasing P450 turnover numbers in intact

bacteria. Another approach to this goal would be overexpression of the endogenous bacterial P450 reductase system. We have purified this system finding it to consist of two components; a soluble FMN-containing protein, flavodoxin, which binds to P450 and an FAD-containing protein, flavodoxin reductase, which reduces flavodoxin using electrons from NADPH (C. Jenkins and M. Waterman, unpublished). Having characterized these two enzymes, a next step is to clone them and overexpress these proteins in bacteria in the hope of enhancing the P450 turnover numbers. Perhaps turnover numbers greater than those obtained with fusion proteins can be attained by this approach. At present the limitation on the use of bacteria as commercial P450 containing bioreactors is the intrinsically low P450 turnover number. We are confident, however, that this limitation will be overcome by one or more of the approaches noted.

Another important consideration for the use of bacterially expressed P450s in the study of their enzymatic properties is the ability to obtain highly purified samples of these enzymes. This is particularly important for the study of drug metabolism by individual forms of human P450 where the use of relatively large amounts of purified P450 in standard reconstitution assays will be important for producing quantities of metabolites from a test drug sufficient for structural analysis. Microsomal P450s have been purified easily from bacterial membranes by conventional purification techniques. Also these enzymes can be purified by attaching affinity tags to them and using affinity chromatography after solubilizing the bacterial membranes. Successful use of this latter approach has involved addition of four histidine residues at the carboxy-terminus of P450s and purification by nickel ion chromatography (Imai et al. 1993). Such histidine residues have no effect on the enzymatic activity, and for the one reported example 56 mg homogenous human P450c17 was obtained from 9 l bacterial culture. Thus, a previously inaccessible human P450 has been obtained in large quantities by bacterial expression.

5.5 Use of Bacterial Expression To Study P450 Structure/Function

As a tool to understanding the biochemistry of P450s leading to production of engineered hydroxylases (designer P450s) the bacterial expression system has important advantages over other expression systems. The first and most obvious is high-level expression. As described
in Sect. 5.4, large amounts of difficult to obtain forms of P450 can be
obtained with relative ease using the bacterial expression system. Thus
one can study not only the enzymatic properties but also the biophysical
properties of P450s. Other expression systems can be used to study
enzymatic properties, but often these systems cannot be used for biophysical studies due to the low level of expression. Second, bacterial
systems are very facile for site-directed mutagenesis studies. To date,
this expression system has been most effectively used in mutagenesis
studies of the mitochondrial P450, P450scc.

Based on chemical modification studies and protein sequence alignment it was predicted that two specific lysine residues in bovine
P450scc might be important in binding adrenodoxin. Mutation of these
residues (Lys-338 and Lys-342) to neutral or acidic amino acids was
found to have a profound effect on the K values of adrenodoxin binding
(Table 2) without affecting the enzymatic properties (pregnenolone production from 25-hydroxycholesterol; Wada and Waterman 1992). To
carry out these studies it was necessary to purify recombinant P450scc
from *E. coli* and to measure optically the binding of adrenodoxin to
P450scc by titration. The high level of expression of P450scc in bacteria
made these experiments possible. In a more recent study site-directed
mutagenesis has been used to probe the active site of P450scc (I.
Pikuleva and M. Waterman, unpublished). Tyrosine residues 93 and 94
align closely with Tyr-96 of P450cam and thus are predicted to play a
role in substrate binding in P450scc. These particular tyrosine residues
are highly conserved among P450scc from several species, further supporting this notion. One of the serious problems associated with investigation of protein structure/function relationships by site-directed mutagenesis was encountered in these particular studies, that being the
effect of amino acid alterations on protein stability. Thus the single
mutants (Y93S, Y93A, Y94S, Y94A) were not very stable, and only
limited quantities of these P450scc mutants could be obtained during

Table 2. Estimated binding constants
of adrenodoxin to P450scc

Wild type	0.23 μM
Lys-338 Ala	60.0 μM
Lys-338 Gln	74.0 μM
Lys-338 Thr	130.0 μM
Lys-338 Glu	130.0 μM
Lys-338 Glu	130.0 μM
Lys-342 Glu	130.0 μM
Lys-342 Gln	35.0 μM

purification. However, a functional double mutant (Y93S, Y94S) was obtained in sufficient quantities for optical measurement of substrate binding constants by titration. This double mutant bound cholesterol five times less well than did the wild type enzyme, while it bound 22-hydroxycholesterol, the first intermediate in the conversion of cholesterol to pregnenolone, with the same affinity as the wild-type enzyme. It is concluded that one or both of these tyrosine residues is important for cholesterol binding but is not involved in 22-hydroxycholesterol binding. Studies of enzymatic properties of the single mutants indicate that the key residue is Y93. Once again, the relatively high level of expression of P450scc in bacteria made this study possible.

We are rapidly becoming convinced, however, that site-directed mutagenesis resembles somewhat Russian roulette because many mutations based on the most careful alignment of P450 sequences and inspection of known P450 structures do not lead to useful information concerning with the structure or function of the wild-type P450. Such mutations either have no effect on activity or completely alter the structure of the P450, leading to P420. Thus there is a strong need for X-ray structures of eukaryotic P450s, both mitochondrial and microsomal forms. Without such information the production of engineered designer P450s will remain impossible. The bacterial expression system makes it possible to produce large quantities of P450s for crystallization attempts. As more laboratories participate in the effort to obtain the structure of eukaryotic P450s, the closer we come to achieving that goal.

5.6 Disadvantages of Bacterial Expression of P450s

In the above discussion disadvantages of the bacterial expression system for P450 investigation have been noted, and the purpose in this section is simply to summarize them. In many cases modification of the amino-terminal sequences has been used to achieve high level expression, and even if only one amino acid is changed, it raises the question in some minds as to whether the recombinant P450s provide an accurate and precise representation of the activity of the native enzyme. It should be noted here that in no case have such minor changes in amino acid sequences been found to alter the enzymatic activities. Second, bacteria do not naturally contain reductase systems which support high level P450 activities. Thus purification or construction of fusion proteins seems necessary to study P450 enzymatic activities. Our view is that these disadvantages are more than offset by the ability to obtain large amounts of difficult to obtain human enzymes. Third, the bacterial system is unpredictable. As noted above, high-level expression of P450s in bacteria is strain dependent for unknown reasons. Also, we have found that the folding pathway for P450s in mammalian cells and bacteria can be quite different. Removing the amino terminal signal anchor from microsomal P450s leads to an incorrectly folded nonfunctional enzyme in mammalian cells presumably due to the necessary role of the signal recognition particle pathway in P450 folding (Clark and Waterman 1992). When such a truncated form of P450 is expressed in *E. coli* however, a functional enzyme is obtained suggesting that the P450 folding pathway in bacteria does not involve the signal recognition particle pathway (Sagara et al. 1993). The effects of such differences between mammalian and bacterial cells on the final P450 product are not known and based on enzymatic activity measurements not important. Nevertheless, such variability is unsettling. Finally it is unclear how well bacterial membranes resemble the endoplasmic reticulum in their ability to participate in folding P450s and in their enzymatic activities. This latter concern can also be raised when considering yeast or insects as expression systems. Briefly stated, there are differences between *E. coli* and human cells. Whether such differences lead to variations in P450 structure and/or function are unknown. No such variations have yet been documented. Our view is that the ability to obtain very large quantities of P450s previously not available outweighs such concerns.

Table 3. Expression of bovine P-450c17 in different expression system: relative levels determined by imunnoblot analysis from 25 μg membrane protein

Expression system	Relative level
Bovine adrenal microsomes	1.0
COS microsomes	0.25
Yeast microsomes	0.8
Sf9 (insect) microsomes	2.0
E. coli membranes	3.0

5.7 Conclusions

While the bacterial expression system has proved to work well for the expression of functional forms of both microsomal and of mitochondrial P450, we understand the requirements for their expression in bacteria far less clearly than those for expression of P450s in eukaryotic cells. Nevertheless, the ability to generate large quantities of P450 (Table 3) in an inexpensive culture system, the ease of site-directed mutagenesis in bacteria and the ability to easily purify relatively large quantities of P450s, including human enzymes that are not normally available, make this a particularly useful expression system for P450 structure/function analysis. Furthermore, the background level of P450s in E. coli is extremely low or nonexistent.

Of all the expression systems used for P450 research, the bacterial system seems to hold the greatest promise for providing useful bioreactors that catalyze engineered chemical reactions. It can be predicted that, either by overexpression of the unknown, endogenous bacterial reductase system or by the formation of P450/P450 reductase fusion proteins, bioreactors with high levels of engineered P450 activities for commercial use will be produced.

Acknowledgments. The support of USPHS grant GM37942 is greatly appreciated. The author appreciates contributions to this work by Drs. Henry Barnes, Akira Wada, Tsuneo Imai, Takujii Sawaya, Norio Kagawa, Irina Pikuleva, and Chris Jenkins.

References

Asseffa A, Smith SJ, Nagata K, Gillette J, Gelboin HV, Gonzalez FJ (1989) Novel exogenous heme-dependent expression of mammalian cytochrome P450 using baculovirus. Arch Biochem Biophys 274:481–490

Barnes HJ, Arlotto MP, Waterman MR (1991) Expression and enzymatic activity of recombinant cytochrome P450 17α-hydroxylase in escherichia coli. Proc Natl Acad Sci USA 88:5597–5601

Clark BJ, Waterman MR (1992) Functional expression of bovine 17α-hydroxylase in COS 1 cells is dependent upon the presence of an amino-terminal signal anchor sequence. J Biol Chem 267:24568–24574

Crespi CL (1991) Expression of cytochrome P450 cDNAs in human B lymphoblastoid cells: applications to toxicology and metabolite analysis. In: Waterman MR, Johnson EF (eds) Methods in enzymology, vol 206. Academic, New York, pp 123–129

Doehmer J, Oesch F (1991) V79 Chinese hamster cells genetically engineered for stable expression of cytochromes P450. In: Waterman MR, Johnson EF (eds) Methods in enzymology, vol 206. Academic, New York, pp 117–122

DuBois RN, Simpson ER, Tuckey J, Lambeth JD, Waterman MR (1981) Evidence for a higher molecular weight precursor of cholesterol side-chain-cleavage cytochrome P-450 and induction of mitochondrial and cytosolic proteins by corticotropin in adult bovine adrenal cells. Proc Natl Acad Sci U S A 78:1028–1032

Fisher CW, Shet MS, Caudle DL, Martin-Wixtrom CA, Estabrook RW (1992) High-level expression in escherichia coli of enzymatically active fusion proteins containing the domains of mammalian cytochromes P450 and NADPH-P450 reductase flavoprotein. Proc Natl Acad Sci USA 89:10817–10821

Gonzalez FJ, Aoyama T, Gelboin HV (1991) Expression of mammalian cytochrome P450 using vaccinia virus. In: Waterman MR, Johnson EF (eds) Methods in enzymology, vol 206. Academic, New York, pp 85–92

Imai T, Globerman H, Gertner JM, Kagawa N, Waterman MR (1993) Expression and purification of functional human 17α-hydroxylase/17,20-lyase (P45017a) in Escherichia coli: use of this system for study of a novel form of combined 17α-hydroxylase/17,20-lyase deficiency. J Biol Chem 268:19681–19689

Larson JR, Coon MJ, Porter TD (1991) Alcohol-inducible cytochrome P45011E1 lacking the hydrophobic NH2-terminal segment retains catalytic activity and is membrane-bound when expressed in Escherichia coli. J Biol Chem 266:7321–7324

Li Y-C, Chiang JYL (1991) The expression of a catalytically active cholesterol 17α-hydroxylase cytochrome P450 in Escherichia coli. J Biol Chem 266:19186–19191

Lu AY, Junk KW, Coon MJ (1969) Resolution of the cytochrome P450-containing ϖ-hydroxylation system of liver microsomes into three components. J Biol Chem 244:3714–3721

Narhi LO, Fulco AJ (1987) Identification and characterization of two functional domains in cytochrome P-450$_{BM-3}$', a catalytically self-sufficient monooxygenase induced by barbiturates in bacillus megaterium. J Biol Chem 262:6683–6690

Oeda K, Sahaki T, Ohkawa H (1985) Expression of rat liver cytochrome P450MC cDNA in saccharomyces cerevisiae. DNA 4:203–210

Porter TD, Wilson TE, Kasper CB (1987) Expression of a functional 78,000 dalton mammalian flavoprotein, NADPH-cytochrome P-450 oxidoreductase, in Escherichia coli. Biochem Biophys 254:353–367

Sagara Y, Barnes HJ, Waterman MR (1993) Expression in Escherichia coli of functional cytochrome P450c17 lacking its hydrophobic amino-terminal signal anchor. Arch Biochem Biophys 304:372–378

Wada A, Waterman MR (1992) Identification by site-directed mutagenesis of two lysine residues in cholesterol side chain cleavage cytochrome P450 that are essential for adrenodoxin binding. J Biol Chem 267:22877–22882

Wada A, Matthew PA, Barnes HJ, Sanders D, Estabrook RW, Waterman MR (1991) Expression of functional bovine cholesterol side chain cleavage cytochrome P450 (P450scc) in Escherichia coli. Arch Biochem Biophys 290:376–380

Waterman MR (1994) Heterologous expression of mammalian P450 enzymes. Adv Enzymol 68:37–66

Yabusaki Y, Murakami H, Sakaki T, Shibata M, Ohkawa H (1988) Genetically engineered modification of P450 monooxygenases: functional analysis of the amino-terminal hydrophobic region and hinge region of the P450/reductase fused enzyme. DNA 7:701–711

Zuber MX, Simpson ER, Waterman MR (1986) Expression of bovine 17α-hydroxylase cytochrome P-450 cDNA in nonsteroidogenic (COS 1) cells. Science 234:1258–1261

6 Expression of Cytochromes P450 in Yeast: Practical Aspects

D. Pompon, G. Truan, A. Bellamine, and P. Urban

6.1   Introduction 97
6.2   The Yeast Background 99
6.3   Basic Expression of Heterologous P450s in Yeast 100
6.4   Tuning the Yeast Redox Environment To Optimize Heterologous
        P450 Activity.. 103
6.5   Adding Phase II Activity at Will........................ 107
6.6   Conclusions 108
References 109

6.1 Introduction

Heterologous expression systems must ideally feature high, stable, reproducible and low-cost synthesis of any P450 under a native folding state, including full saturation by the hemin prosthetic group. Additionally, the host cell must be free of any interfering endogenous P450 activity and must offer a suitable P450 environment mimicking as closely as possible the natural microsomal membrane of the organism from which the P450 originates. Production of large amounts of metabolites further requires the reconstitution of an in vivo self-sufficient system. The natural presence or the coexpression of P450-associated electron transfer proteins, particularly NADPH-dependent reductase and cytochrome b_5, is thus useful. In contrast, the presence of endogenous phase II activities in the host system could be deleterious because it sometimes results in conjugation reactions that mask the

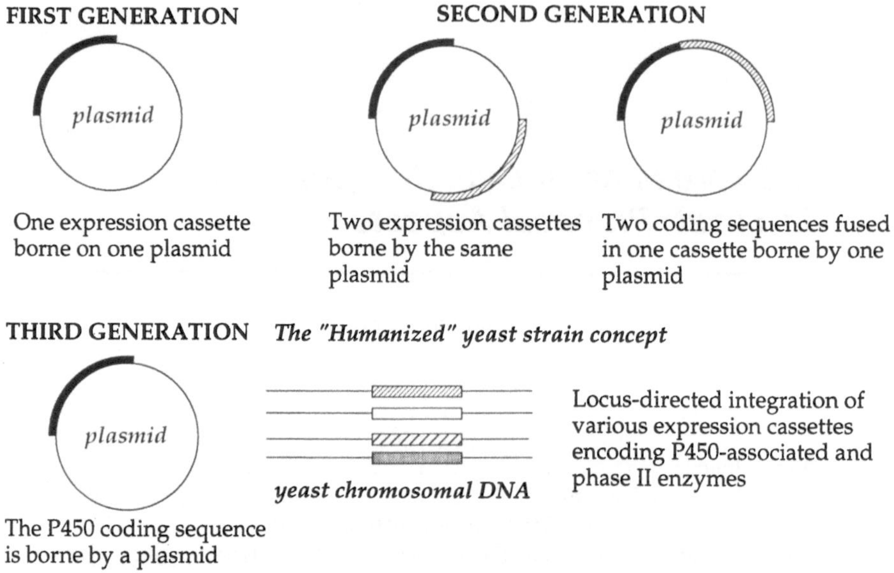

FIRST GENERATION   **SECOND GENERATION**

One expression cassette   Two expression cassettes   Two coding sequences fused
borne on one plasmid   borne by the same   in one cassette borne by one
                                 plasmid   plasmid

THIRD GENERATION   *The "Humanized" yeast strain concept*

Locus-directed integration of
various expression cassettes
encoding P450-associated and
phase II enzymes

yeast chromosomal DNA

The P450 coding sequence
is borne by a plasmid

Fig. 1. The different generations of P450-expressing yeast systems

formation of reactive intermediates. Nevertheless, some P450-dependent activation reactions involve an obligatory association with a phase II enzyme, such as microsomal epoxide hydrolase, to generate the intermediates then used as substrates by the same or other P450 to produce the final activated metabolites. Optional coexpression of phase II activity is thus an interesting feature for a more accurate simulation of drug and pollutant metabolism. Among the numerous organisms in which expression tools are available, yeast is rather unique in meeting all the different criteria previously listed.

Therefore, since the first report on the expression in yeast of a mammalian P450 (Oeda et al. 1985) *Saccharomyces cerevisix* has become one of the most popular systems for heterologous expression of P450-dependent mono-oxygenase activities (Urban et al. 1990; Yabusaki and Ohkawa 1991). The principal purpose of this contribution is to present the different generations of yeast expression systems that succeeded one the other during the past 8 years, leading from a simple expression system to the concept of "humanized" cells (Fig. 1).

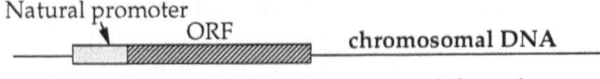

A natural yeast gene with its promoter and the coding
sequence

FIRST TYPE: PROMOTER SUBSTITUTION

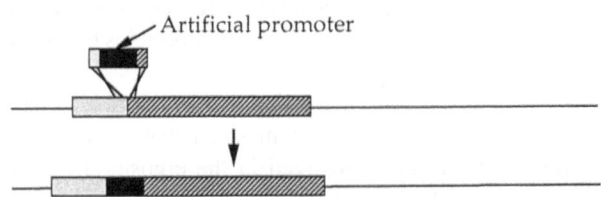

The resultant strain presents a modified regulation for the
engineered locus

SECOND TYPE: GENE REPLACEMENT

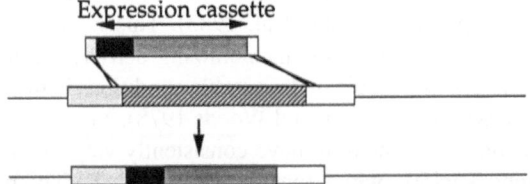

The resultant engineered strain presents a disrupted-like
phenotype for the targeted locus and harbors a new gene

Fig. 2. Gene engineering in yeast

6.2 Yeast Background

The yeast *S. cerevisiae* (baker's yeast) is a eukaryotic micro-organism
featuring a subcellular membrane organization very similar to that of
any higher eucaryotic cell. Its endoplasmic reticulum membrane contains endogenous NADPH-P450 reductase (about 20 pmol/mg) and significant amounts of cytochrome b_5 (about 80 pmol/mg). A rather efficient redox coupling between human P450s and the yeast P450
reductase has been found to occur, although human and yeast P450
reductases exhibit only 33% amino acid sequence identity. The situation

appears quite different for the coupling with cytochrome b_5. Although yeast and human cytochromes b_5 share approximately 30% sequence identity (Truan et al. 1994), no detectable coupling was observed when yeast cytochrome b_5 was tested for supporting human P450 3A4-catalyzed activity (Truan et al. 1993). Fortunately, the availability of a wide range of molecular biological tools and the particular recombination features of yeast cells make clean gene substitutions easy, including the exchange of endogenous P450 associated redox proteins by their human equivalents (Fig. 2).

    S. cerevisiae cells contains, according to the present understanding, three endogenous P450s involved in domestic metabolisms. Lanosterol-C14-demethylase (CYP51) is involved in the ergosterol biosynthesis pathway (Kalb et al. 1987), and the DIT2 gene product (CYP56), a dityrosine synthase, is expressed only during sporulation phase (Briza et al. 1990). These P450s seem almost or totally inactive toward xenobiotic compounds. The third P450, still putative, has been postulated to support the A-22 desaturation of the side chain of the ergosta-5,7-dienol, an ergosterol precursor (Hata et al. 1983). This isoform or some other could have some xenobiotic-metabolizing activity (Kelly et al. 1985) but is almost undetectable under aerobic conditions in the absence of catabolic repression (Wiseman and Woods 1978). Yeast cells grown in a galactose-containing medium have consistently very low amounts of total endogenous P450s, which remain undetectable both on the basis of spectral analysis and of the assay of xenobiotic-directed activities.

6.3 Basic Expression of Heterologous P450s in Yeast

Expression in yeast of human P450s is based on the construction of expression cassettes containing the open reading frame (ORF) of the considered human cDNA sandwiched between a yeast transcription promoter *(PGK, GPD, ADH1, PHO5, GALIO,...)* and termination sequences (Fig. 3A). Total poly(A)-rich mRNAs are prepared, and cDNAs are reverse-transcribed. Polymerase chain reaction (PCR) amplification is performed using a pair of specific primers and total cDNAs as template. Each primer is composed of two parts, its 5'-end bears a restriction site compatible with the ones of the expression vector, and its 3'-end is identical (PCR 5'-primer) or complementary (PCR 3' primer)

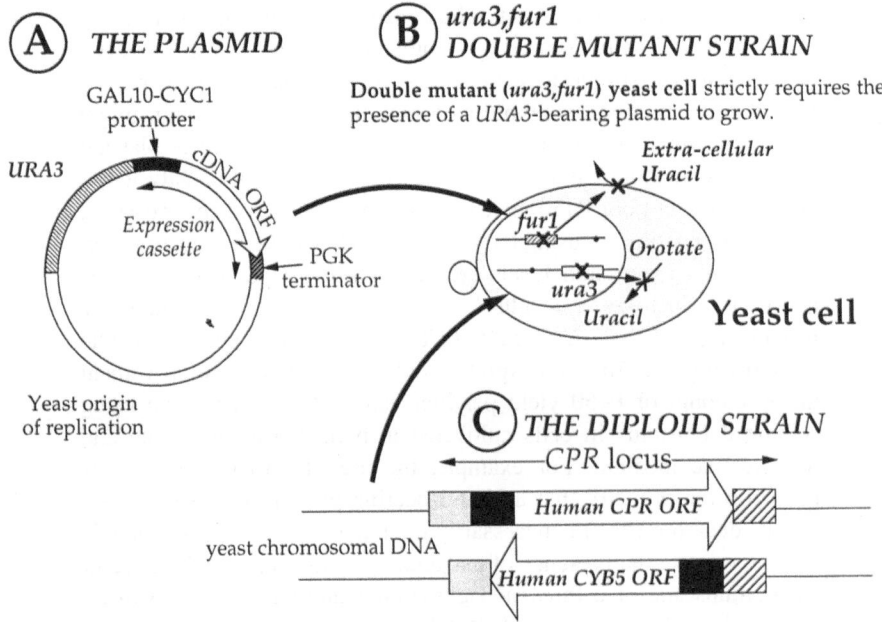

Fig. 3A–C. Procedures for humanizing yeast

to 18–25 bp of the template sequence. Particular attention must be paid in designing primers to remove the totality of the cDNA 5'-non-coding sequence if optimal expression is to be achieved (Pompon 1988). When it occurs, any hairpin structure surrounding the initiation codon, as is the case with some human mRNAs and in particular that encoding human P450 reductase (Yamano et al. 1989), must be eliminated by introducing suitable mutations into the PCR 5'-primer sequence. Indeed, transcription in yeast as well as translation is strongly inhibited by such hairpin structures (Baim and Sherman 1988). After sequencing, the PCR-amplified ORF-containing fragment is transferred into a multicopy yeast expression vector bearing suitable auxotrophy complementation markers. The multicopy self-replicating plasmids involved are simply introduced in the suitable yeast cell and selected on the basis of the marker complementation.

Systems involving plasmids with constitutive promotors and culture at the low cell density allowed by commercial minimum media (2×10^7 cells/ml) are the simplest for those who are unexperienced. Nevertheless, the constitutive overexpression of a heterologous protein might cause a negative selection pressure reading to a low average plasmid copy number. This feature associated with the low cell density led to generally rather poor P450 yields (1–10 nmol/l), making spectroscopic studies or experiments using large-scale of P450 out of range. However, even limited expression levels are convenient for addressing simple questions. Much better results are generally obtained with inducible promoters *(GALT, GAL10CYC1, PHO5)* which allow one to separate the growth phase from the expression phase. A dramatic improvement in the amount of P450 yield per liter can be further achieved using stabilized plasmids in cells cultivated at high density in a rich and nonselective medium. For example, the use of host-vector systems based on the property that a URA3-bearing plasmid (URA3 encodes orotate decarboxylase) is necessarily maintained in a *ura3, furl* double mutant, even in the presence of exogenous uracil (Fig. 3B). The use of such strains and of a three-step glucose-ethanol-galactose growth-induction procedure (Truan et al. 1993), which allows the full separation of the biomass growth phase from the expression phase, allowed us to reach cell densities up to two orders of magnitude higher than that observed in basic expression systems. P450 yields exceeding 500 nmol/l and specific content of 200–400 pmol/mg microsomal protein were obtained. These values compare fairly well with the best results obtained in *Escherichia coli* for modified P450s carrying the signal sequence of bovine P450 17α-hydroxylase (Barnes et al. 1991; Fisher et al. 1992).

Heterologous P450s produced in *S. cerevisiae* cells are generally obtained in the natural folded state without any particular precautions. For yeast subcellular fractionation, however, a critical step is the breaking of the cellular wall without damaging P450 molecules. Being an aggregate of mannoproteins, complex sugars and chitin, the cell wall is quite rigid. Two methods, based respectively on mechanical disruption and on enzymatic digestion of the cell wall with endoglucanases, are used. Mechanical breaking is preferred for large-scale culture at high cell density, whereas the enzymic treatment is more suitable for small-scale culture, particularly when contaminations by mitochondrial fractions must be avoided. In contrast to *Escherichia coli*, the presence in yeast of a significant amount of endogenous P450 reductase gives a self-sufficient system for the monitoring of P450-catalyzed activity without needing in vitro addition of any additional component. Therefore, whole cells or microsomes prepared from P450-expressing yeast are a material of choice for activity screening purposes or structure-function studies. Nevertheless, the turnover of P450s produced in wild-type strains is often significantly lower than that determined after purification and reconstitution with an excess of mammalian P450 reductase. The rationale was found in the limiting amount of yeast P450 reductase (Urban et al. 1990; Murakami et al. 1990). As a demonstration, the turnover for mouse P450 1A1 ethoxyresorufin-*O*-deethylase activity analyzed in yeast microsomes during the course of the P450 induction was found effectively to decrease progressively in parallel with increasing P450 accumulation (Urban et al. 1990). This suggested that endogenous NADPH-P450 reductase, at the low concentration found in native yeast, is rather inefficient in supporting P450 activity at full.

6.4 Tuning the Yeast Redox Environment To Optimize Heterologous P450 Activity

Although the endogenous P450 reductase content in yeast microsomes is, when expressed as a NADPH-cytochrome *c* oxidoreductase activity, similar to the values found in liver, the heterologous nature of the P450 reductase – P450 couple results in a rather inefficient coupling. A first solution considered was to increase significantly the endogenous P450 reductase activity by its overexpression, either from a multicopy vector (Urban et al. 1990; Murakami et al. 1990) or from a yeast strain engineered at the genomic DNA level (Truan et al. 1993). The increase in P450 activity by overexpression of yeast P450 reductase from a second plasmid is hardly possible due to the known genetic instability of such cotransformants. We did not consider alternate solutions involving construction of a single plasmid containing several expression or construction of a fusion protein between yeast P450 reductase and the P450 of interest (Murakami et al. 1986; Shibata et al. 1990). Particularly, construction of artificial fusion protein could led to unpredictable alteration

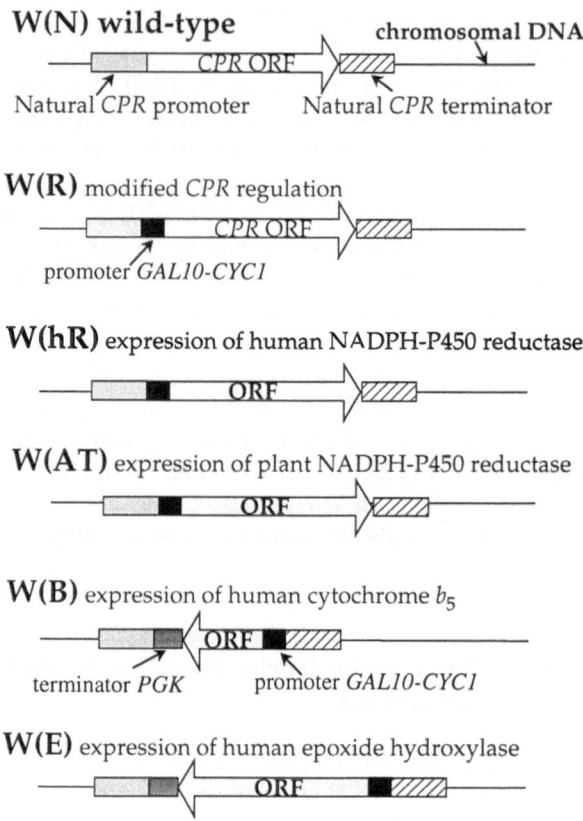

Fig. 4. Structures of the *CPR* locus in the engineered yeast strains

of the P450 substrate specificity. Finally, the genomic integration of a yeast P450 reductase overexpression cassette was retained as the sole solution reading to a stable coexpression without any alteration of the heterologous P450 produced (Figs. 3C, 4).

W(R) strain which overexpresses the yeast P450 reductase was constructed from the wild-type W(N) strain (W303-1B) by replacing the natural promoter of the endogenous *CPR* gene (which encodes the yeast NADPH-P450 reductase) by an artificial galactose-inducible one (Truan et al. 1993). When grown on glucose, W(R) cells do not express detectable levels of P450 reductase; thus behaving as a *CPR*-disrupted strain. However, when grown on galactose, the yeast P450 reductase reached in approximately 10 h a specific microsomal activity 30-fold higher than the value measured in a wild-type strain. This increased P450 reductase activity was found dramatically to enhance heterologous P450 turnover, although to various extents depending on the P450 isoform (Truan et al. 1993). For instance, human and mouse CYP1Als exhibit respectively a five- and tenfold increased turnover for ethoxyres-orufin-*O*-deethylase activity compared to that in W(N) wild-type yeast. Contrasting with this moderate effect, human CYP3A4 and jerusalem artichoke CYP73 exhibit, respectively, 62- and 21-fold higher turnover for testosterone 6β-hydroxylase and cinnamate 4-hydroxylase activities (Truan et al. 1993; Urban et al. 1994). These results demonstrate that overexpression of yeast P450 reductase permits the significant enhancement of turnover of mammalian and plant P450s produced in yeast. High P450 reductase levels compensate for the rather inefficient coupling between heterologous partners. However, the strong overexpression of the yeast P450 reductase results in generation of higher and toxic amounts of dioxygen-derived radicals, possibly resulting in membrane or P450 damages. Moreover, artifactual xenobiotic biotransformations could result from direct redox reactions with the overexpressed P450 reductase, bypassing the heterologous P450 when the W(R) system is used for drug metabolism prediction.

Clean substitution of the endogenous P450 reductase by heterologous ones to reconstitute homologous P450 reductase – P450 couples was consequently tested, assuming that lower amounts of a P450 reductase homologous to the expressed P450 would be required to support P450 activity as efficiently as observed in W(R) cells. The approach consisted in the substitution of both the promoter and the ORF of the yeast genomic *CPR* gene by the ones of a P450 reductase-encoding gene originating from the plant *A. thaliana* (resulting in WAT11 and WAT22 yeast strains) or from human [yielding W(hR) yeast strain] (Urban et al. 1993; Pompon et al. 1994; Fig. 4). In the case of the human P450 reductase we determined that elimination of an hairpin structure standing in the 5'-end of human P450 reductase-encoding cDNA was needed to obtain appreciable human P450 reductase expression in yeast (Urban et al. 1993). The "humanized" W(hR), and the "vegetalized" WAT11 or WAT22 strains were found perfectly viable although not expressing yeast P450 reductase (Pompon et al. 1994). These engineered strains offer an improved redox environment for mammalian and plant P450s expressed from a plasmid. Despite that the P450 1A1-catalyzed bioconversion rate was found similar in W(hR) cells transformed by human or mouse CYP1A1 compared to W(R) transformed cells, the human P450 reductase content of W(hR) yeast microsomes is found to be 90-fold lower than that in W(R) (Urban et al. 1993). These preliminary results indicate that the human P450 reductase expressed in yeast seems to be much more efficient than the yeast enzyme in supporting the activity of mammalian P450s, even if its cytochrome c reductase specific activity is much lower.

The effect on P450 activity of cytochrome b_5, a second redox protein involved in electron transfer to P450, was also considered. The endogenous *CPR* gene was replaced in an haploid strain by an expression cassette for human cytochrome b_5, giving W(B) engineered strain. In galactose, W(B) cells produce both yeast and human cytochromes b_5 in similar amounts but no P450 reductase (Truan et al. 1993). The W(B,R) diploid strains was constructed from W(B) and W(R) strains by matting (Fig. 3C). This strain produces both high levels of yeast P450 reductase and human cytochrome b_5 galactose-containing medium. When mating W(hR) with W(B), the resulting strain expresses no P450 reductase when grown on glucose but only human P450 reductase and cytochrome b_5 in galactose.

Table 1 summarizes the effect of increasing the level of yeast P450 reductase activity and/or coexpressing human cytochrome b_5 on yeast-expressed human P450s 1A1 and 3A4, respectively, for ethoxyresorufin-*O*-deethylase (EROD) and testosterone 6β-hydroxylase (THL) activities. The effect is clearly isoenzyme dependent. Human cytochrome b_5 was found to have no or little potentiating effect on P450 1A1 EROD turnover at low P450 reductase activity since a twofold decrease in P450 reductase activity results in a twofold decrease in EROD turnover [compare results in W(B,N) and W(N)]. On the other hand, human cytochrome b_5 was found to strongly potentiate human P450 3A4 since, at low P450 reductase activity, the THL turnover is increased sevenfold in W(B,N) compared to that found in W(N) although P450 reductase activity is halved. This indicates a critical role of human cytochrome b_5 for the THL activity catalyzed by human P450 3A4 at low P450 reductase level. Moreover, yeast cytochrome b_5 does not substitute for human

Table 1. Turnover numbers of human P450s 1A1 and 3A4 expressed in engineered strains

Strains	NADPH P450 reductase (nmol mg/min)	Human cytochromes b_5	Human P450 1A1 EROD turnover (min^{-1})	Human P450 3A4 THL turnover (min^{-1})
W(N)	100	No	3.2	0.03
W(R)	1800	No	14.0	1.92
W(B,R)	800	Yes	16.0	2.26
W(B,N)	50	Yes	1.6	0.22
W(R,N)	800	No	11.2	1.36

cytochrome b_5, since this yeast protein is produced in all strains and does not mask the effect of human cytochrome b_5. Comparison of 3A4-catalyzed THL activity in W(N) and W(R) cells indicates indeed that a large overproduction of the yeast P450 reductase seems to compensate for the absence of human cytochrome b_5. As a consequence, the association of human cytochrome b_5 production and yeast P450 reductase overproduction in W(B,R) resulted in a 73-fold increase of P450 3A4-catalyzed THL activity compared to that observed in wild-type yeast W(N). Comparison with results obtained with P450 1A1 suggests that the cytochrome b_5 effect is highly P450 isoenzyme dependent. Complementary results with yeast-expressed P450 3A4 indicated that the cytochrome b_5 effect is also substrate dependent (Peyronneau et al. 1992). We conclude that turnover of human P450 is differentially affected by cytochrome b_5 expression, and that coexpression of human cytochrome b_5 is required for an accurate simulation of P450 activities.

6.5 Adding Phase II Activity at Will

Once yeast strains engineered to express human or plant P450 reductase had been constructed, the next step was to build yeast coexpression systems including, at will, important phase II enzyme such as microsomal epoxide hydrolase. The principle is based on an extension of the previous concept (depicted in Fig. 2) and involves multiple genomic integrations in diploid strains. The expression cassette coding for human microsomal epoxide hydrolase was integrated into the yeast genome using the same procedure as used for W(B) construction, yielding the W(E) strain (Gautier et al. 1993; Fig. 4). Mating of W(E) strain with W(R) strain resulted in the W(E,R) diploid strain which, when grown on galactose, both expresses human epoxide hydrolase, evidenced by the typical styrene-oxide hydrolase activity, and overproduces yeast P450 reductase. Microsomes prepared from W(E,R) cells transformed by a plasmid encoding human P450 1A1 were incubated with benzo[a]pyrene. This led to the efficient production in yeast of the expected products of the human benzo[a]pyrene metabolism, and in particular the 7,8- and 4,5-diol products which are not formed in transformed W(R) strain (Gautier et al. 1993). Interestingly, the formation of the hydrolysis product of the ultimate mutagen (the 7,8,9,10-tetrol) was observed only in the strain coexpressing the yeast P450 reductase, human P450 1A1, and epoxide hydrolase (Gautier et al., manuscript in preparation). The kinetic analysis clearly indicates that the benzo[a]pyrene-7, 8-oxide intermediate does not accumulate, meaning that a very efficient coupling has been established between phase I and phase II human enzymes in yeast microsomes. A three-step metabolic pathway, reading to a genotoxic activation involving phase I and phase II enzymes, was thus introduced in engineered yeast.

6.6 Conclusions

"Humanized" strains offer a new powerful approach for the simulation of single or multistep drug and pollutant metabolisms. The expression of virtually any human or plant P450 in an optimized environment without the need for particular vector construction can be conveniently performed. The only requirement is the transformation by a suitable P450 expression vector of the best-adapted "humanized" or "vege-talized" strain. The multi-integration strategy could be very easily extended to the coexpression of a large number of heterologous enzymes, thus opening the field to simulation and analysis of complex metabolite pathways and to the development of sophisticated biotechnological tools for industrial bioconversions.

References

Baim SB, Sherman F (1988) MRNA structures influencing translation in the yeast Saccharomyces cerevisiae. Mol Cell Biol 8:1591–1601

Barnes HJ, Arlotto MP, Waterman MR (1991) Expression and enzymatic activity of recombinant cytochrome P450 17α-hydroxylase in Escherichia coli. Proc Natl Acad Sci USA 88:5597–5601

Briza P, Breitenbach M, Ellinger A, Segal J (1990) Isolation of two developmentally regulated genes involved in spore wall maturation in Saccharomyces cerevisiae. Gene Dev 4:1775–1789

Fisher CW, Caudle DL, Martin-Wixtron C, Quattrochi LC, Tukey RH, Waterman MR, Estabrook RW (1992) High-level expression of a functional human cytochrome P450 1A2 in Escherichia coli. FASEB J 6:759–764

Gauner JC, Urban P, Beaune P, Pompon D (1993) Engineered yeast cells as model to study coupling between human metabolizing enzymes. Eur J Biochem 211:63–72

Hata S, Nishino T, Katsuki H, Aoyama Y, Yoshida Y (1983) Two species of cytochrome P450 involved in ergosterol biosynthesis of yeast. Biochem Biophys Res Commun 116:162–166

Kalb VF, Woods CW, Turi TG, Dey CR, Sutter TR, Loper JC (1987) Primary structure of the P450 lanosterol demethylase gene from Saccharomyces cerevisiae. DNA 6:529–537

Kelly SL, Kelly DE, King DJ, Wiseman A (1985) Interaction between yeast cytochrome P450 and chemical carcinogens. Carcinogenesis 6:1321–1325

Murakami H, Yabusaki Y, Ohkawa H (1986) Expression of rat NADPH cytochrome P450 reductase CDNA in Saccharomyces cerevisiae. DNA 5:1–10

Murakarni H, Yabusaki Y, Sakaki T, Shibata M, Ohkawa H (1990) Expression of cloned yeast NADPH-cytochrome P450 reductase gene in Saccharomyces cerevisiae. J Biochem (Tokyo) 108:859–865

Oeda K, Sakaki T, Ohkawa H (1985) Expression of rat liver cytochrome P450MC CDNA in Saccharomyces cerevisiae. DNA 4:203–210

Peyronneau MA, Renaud JP, Truan G, Urban P, Pompon D, Mansuy D (1992) Optimization of yeast-expressed human liver cytochrome P450 3A4 catalytic activities by coexpressing NADPH-cytochrome P450 reductase and cytochrome b5. Eur J Biochem 207:109–116

Pompon D (1988) CDNA cloning and functional expression in yeast Saccharomyces cerevisiae of β-naphtoflavone-induced rabbit liver P450 LM4 and LM6. Eur J Biochem 177:285–293

Pompon D, Truan G, Bellamine A, Kazmaier M, Urban P (1994) Coexpression of mammalian, plant or yeast P450s and P450 reductases in Saccharomyces cerevisiae as cloning and bioconversion tools. In: Lechner MC (ed) Cytochrome P450, biochemistry and biophysics. Elsevier, Amsterdam (in press)

Shibata M, Sakaki T, Yabusaki Y, Murakami H, Ohkawa H (1990) Genetically engineered P450 monooxygenases: construction of bovine P450 c17/yeast reductase fused enzyme. DNA Cell Biol 9:27–36

Truan G, Cullin C, Reisdorf P, Urban P, Pompon D (1993) Enhanced in vivo monooxygenase activties of mammalian P450s in engineered yeast cells producing high levels of NADPH-P450 reductase and cytochrome b5. Gene 125:49–55

Truan G, Epinat JC, Rougeulle C, Cullin C, Pompon D (1994) Cloning and characterisation of a yeast cytochrome b5 gene which suppresses ketoconazole hypersensitivity in NADPH P450 reductase disrupted strain. Gene (in press)

Urban P, Cullin C, Pompon D (1990) Maximizing the expression of mammalian cytochrome P450 monooxygenase activities in yeast cells. Biochimie 72:463–472

Urban P, Truan G, Gauner JC, Pompon D (1993) Xenobiotic metabolism in humanized yeast: engineered yeast cells producing human NADPH-cytochrome P450 reductase, cytochrome b5, epoxide hydrolase and P450s. Biochem Soc Transact 21:1028–1033

Urban P, Werck-Reichhart D, Teutsch H, Durst F, Regnier S, Kazmaier M, Pompon D (1994) Characterization of recombinant plant cinnamate 4-hydroxylase produced in yeast: kinetic and spectral properties of the major plant P450 of the phenylpropanoid pathway. Eur J Biochem (submitted)

Wiseman A, Woods LFJ (1978) Regulation of the biosynthesis of cytochrome P450 in yeast. Biochim Biophys Acta 544:615–623

Yabusaki Y, Ohkawa H (1991) Genetic engineering on cytochrome P-450 monooxygenases. In: Ruckpaul K, Rein H (eds) Frontiers in biotransformations, vol 4. Akademie, Berlin, pp 169–190

Yamano S, Aoyama T, McBride OW, Hardwick JP, Gelboin HV, Gonzalez FJ (1989) Human NADPH-P450 oxidoreductase: complementary DNA cloning, sequence and vaccinia virus-mediated expression and localization of the CYPOR gene to chromosome VII. Mol Pharmacol 35:8388

7 Human Cells as an Expression System for Cytochromes P450

C. L. Crespi, R. Langenbach, H. V. Gelboin, F. J. Gonzalez, and B. W. Penman

7.1	Introduction	111
7.2	Methods	113
7.3	Results/Discussion	114
7.3.1	Expression System Development	114
7.3.2	Development of Cell Fractionation Procedures	120
7.3.3	Development of cDNA-Expressing Cell Panel	121
7.3.4	Current Status of the CYP cDNA Panel of Cell Lines	126
7.3.5	Applications to P450 Form-Specific Metabolism	128
7.4	Future Directions	130
References		131

7.1 Introduction

Cytochromes P450 play a principal role in the metabolism of many drugs, pollutants, and other xenobiotics. The extent of metabolism and the specific pathways of metabolism can influence the safety and/or efficacy of a drug or drug candidate. Because of the central role of cytochrome P450s in drug metabolism and also carcinogenesis, considerable effort has been devoted by many laboratories to the development of model systems to study human P450 metabolism (reviews: Gonzalez 1988; Guengerich 1988; Gonzalez et al. 1991; Langenbach et al. 1992).

One useful approach for analyzing cytochrome P450 mediated metabolism and toxicity is in vitro cDNA expression. A wide variety of model expression systems have been developed utilizing bacteria, yeast, rodent cells, human cells and, more recently, intact organisms (Waterman and Johnson 1991; Komori et al. 1993). Our laboratory has focused on development of human cells which express human P450 (Crespi et al. 1990a,b,c, 1991a,b,c, 1993; Davies et al. 1989; Penman et al. 1993). We have chosen to focus on human cells because of their potentially greater relevance to human toxicity than other mammalian cells or simple eukaryotic or prokaryotic systems. Our laboratory has used human B lymphoblastoid cells because of the relative ease of culture and scale-up of this anchorage independent cell type and the availability of a flexible extrachromosomal vector system (Sugden et al. 1985). The particular cell line used in our laboratory, AHH-1 TK+/– cells (Crespi and Thilly 1984), is a versatile indicator cell line for a variety of in vitro toxicologic end-points. Therefore, P450-mediated metabolism can be readily related to toxic or genotoxic effects at the cellular level.

A goal of our laboratory is to develop a comprehensive panel of cell lines expressing individual human cytochrome P450 cDNAs at high levels. Such a panel is useful for analyzing P450 form-specific metabolism or protoxin activation. In recent years we have made considerable progress towards this goal. The purpose of this contribution is to review and summarize some of our experiences developing this system.

There are two challenges to developing such a panel. The first is the multiplicity of human cytochromes P450 expressed in vivo. There are over 12 distinct cytochrome P450 enzymes expressed in human liver. Each enzyme is a separate project. The second challenge is achieving a stable cell line which contains cytochromes P450 at levels which are adequate for the intended studies.

Based on our experience, one needs at least 1 pmol P450/mg microsomal protein in order to observe toxic effects. In fact, protoxin activation is one of the most sensitive indicators of successful cDNA expression in mammalian cells, often detecting an effect at levels below detection by enzyme assay or western blot. However, at these low expression levels the system has very limited applications to studies of xenobiotic metabolism. A system which is useful for drug metabolism requires at least tenfold higher expression levels, of about 10 pmol P450/mg microsomal protein. At this level primary metabolites of high or moderate turnover substrates are easily detected, particularly if radiolabeled material is used. A system of even greater utility for studying drug metabolism should contain about 50 pmol P450/mg microsomal protein. At this expression level secondary metabolites are readily detected, low turnover substrates can be studied, and metabolism can often be measured by loss of the parent compound. The above expression levels assume adequate but not necessarily saturating cytochrome P450 reductase levels and reasonable linearity of the system (i.e., linear metabolism in microsomes or lysates for about 30 min). If a system is linear for shorter time periods, higher expression levels are needed; if the system is linear for longer periods of time, low expression levels are adequate. The 50 pmol/mg microsomal protein level of expression is roughly comparable to levels of individual P450s in human liver microsomes.

7.2 Methods

Cells, Tissue Culture, DNA Introduction. AHH-1 TK+/- is a human B lymphoblastoid cell line. Cells were maintained in RPMI medium 1640 supplemented to 9% v/v with horse serum. Protoplast fusion has been used to introduce vector into human lymphoblasts and was performed according to Yoakum (1984). Typical stable transfection frequencies are 1/1000 (Crespi 1991). More recently we have used electroporation, which yields stable transfection frequencies of up to 10%. Cells bearing recombinant plasmids were maintained in medium without histidine and containing 2 mM l-histidinol (pEBVHistk-based vectors, Crespi et al. 1990c) or 100–200 µg/ml hygromycin B (pMF6-based vectors; Davies et al. 1989).

Cytotoxicity and Mutagenicity Assays. Cytotoxicity was estimated by measuring growth after treatment. After cultures had resumed exponential growth, the cumulative growth of the mutagen-treated cultures was divided by the cumulative growth of the negative control cultures to yield relative survival. Induction of mutation at the hprt locus was measured by previously published protocols (Penman and Crespi 1987) with minor modifications.

Preparation of Microsomal Samples. Microsomes were prepared according to Penman et al. (1993). For microsomes used in spectral studies, cells were incubated in the presence of 10 μg/ml 5-aminolevulinic acid and 0.2% dimethylsulfoxide (DMSO) for 1 day prior to microsome preparation. Alternatively, 3 μg/ml hemin with 0.2% DMSO was used for routine preparations of microsomes. The use of hemin resulted in preparations which had about 30% higher catalytic activity relative to 5-aminolevulinic acid; however, the microsomes from hemin-treated cells contained a substantial P420 peak. Microsomal protein concentrations were determined by the method of Lowry et al. (1951) using bovine serum albumin as the standard.

Enzyme Assays. 7-Ethoxyresorufin O-deethylase activity was measured in whole cells according to Crespi et al. (1985) and in microsomes according to Burke et al. (1977). 7-Ethoxycoumarin O-deethylase activity and coumarin 7-hydroxylase activity were measured according to Greenlee and Poland (1978). 7-Ethoxy-4-trifluoromethylcoumarin O-deethylase was measured according to DeLuca et al. (1988). Bufuralol 1'-hydroxylase activity was measured according to Kronbach et al. (1987). Microsomes were assayed at 0.5–1 mg protein per milliliter. Testosterone hydroxylase was performed according to Waxman et al. (1983). Cytochrome P450 difference spectra were performed according to Omura and Sato (1964) using and Aminco DW-2000 spectrophotometer.

7.3 Results/Discussion

7.3.1 Expression System Development

The key to the successful development of a panel of human cell lines expressing high level P450 cDNAs has been the development of a flexible and robust expression vector system for use in the cell line. The cell line AHH-1 TK+/- is an immortal human B lymphoblastoid cell line originally isolated from RPMI 1788 cells (Crespi and Thilly 1984). The presumptive immortalizing agent of RPMI 1788 cells is Epstein Barr virus (EBV) which is present in most adult humans. (The AHH-1 TK+/-cells had been tested for the production of infectious virus and found to be negative.) Sugden et al. (1985) developed an extrachromosomal vector which functions in EBV-transformed cell lines. These vectors contain the OriP sequences which function (with the EBNA-1 gene product) as an origin of replication for stable extrachromosomal replication. Our initial studies adapted the pHEBo vector developed by Sugden to introduce a expression cassette for P450 cDNA expression (pMF6, Davies et al. 1989). Over the years we have optimized the expression vector system to achieve higher levels of cDNA expression.

The pHEBo vector and its derivative pMF6 confer resistance to hygromycin B. When hygromycin-resistant cells are selected, the vector is present at a relatively low five copies per cell. Higher vector copy numbers held the promise of increasing cDNA expression through higher "gene dosage". A method to increase vector copy number is to increase the stringency of selection for the vector. Our human lymphoblastoid cells are less sensitive to G418 than to hygromycin B; therefore, G418 selection is unlikely to yield higher vector copy numbers. We developed a means for selection for the vector based on resistance to l-histidinol (Crespi 1992). It turns out that this means of selection is substantially more stringent than hygromycin B. Vector copy numbers were increased from about 5 per cell to about 40 per cell (Crespi et al. 1990c). cDNA expression levels were also increased by a comparable amount. Most of the cell lines which we are currently using use l-histidinol resistance as a means to select for the vector. Hygromycin resistance is now used primarily in the early stages of cDNA evaluation and for applications where multiple cDNAs are to be expressed on individual vectors (Crespi et al. 1991c).

In addition to the vector copy number, another primary determinant of cDNA expression level is the strength of the promoter used for cDNA expression. Our initial adaptation of the pHEBo vector used the herpes simplex virus thymidine kinase gene promoter (HSVtk). This was a conservative choice based on use of that the same promoter to confer resistance to hygromycin B. Subsequent experimentation has revealed that the choice of HSVtk was quite fortuitous. Examination of other promoters including the cytomegalovirus early intermediate and SV40 early region indicated that these promoters give two- and fivefold lower expression levels than HSVtk respectively. These analyses were conducted using a CYP2A6 cDNA and were confirmed using CYP3A4 and CYP2E1 cDNAs.

YEAR

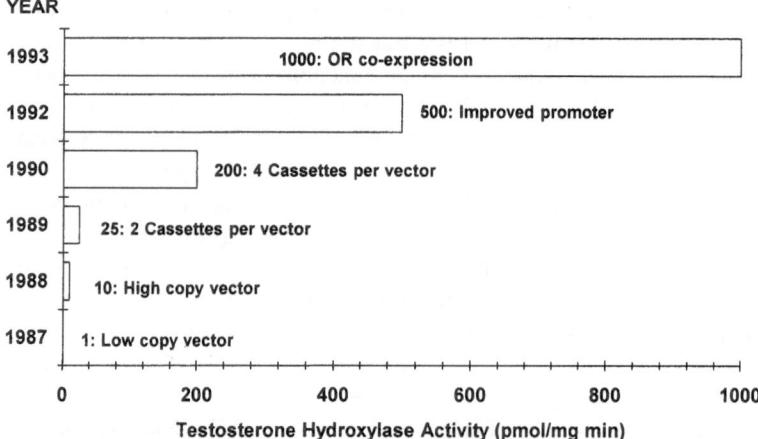

Fig. 1. The microsomal testosterone 6β-hydroxylase activity is plotted for different vector configurations. In 1987 the CYP3A4 cDNA was expressed in the pMF6 vector. In 1988 the higher copy number pEBVHistk vector was used, and then in 1989 and 1990 two and four expression unit were used, respectively. The promoter was modified to allow higher expression with a single expression unit in 1992. Finally, cytochrome P450 reductase cDNA was coexpressed in 1993

Beyond use of the higher copy number vector and the strength available promoter, we have taken additional steps to increase cDNA expression levels and P450 catalytic activity. In particular, we have successfully utilized multiple cDNA-containing P450 expression cassettes per vector molecule (Crespi et al. 1991a, 1993; Penman et al. 1993) and have more recently also coexpressed cytochrome P450 cDNAs with the P450 oxidoreductase (OR) cDNA.

The evolution of cell lines expressing CYP3A4 provides an illustration of the progress which has been made. Figure 1 summarizes the level CYP3A4 expression based on property of the vector system. The year the modification was made is also listed. Initially we obtained an extremely low, barely detectable, CYP3A4 catalyzed testosterone 6β-hydroxylase activity of 1 pmol mg^{-1} min^{-1}. Expression levels were improved by use of a high copy number vector, followed by increasing the number of expression units per vector. The promoter was then modified,

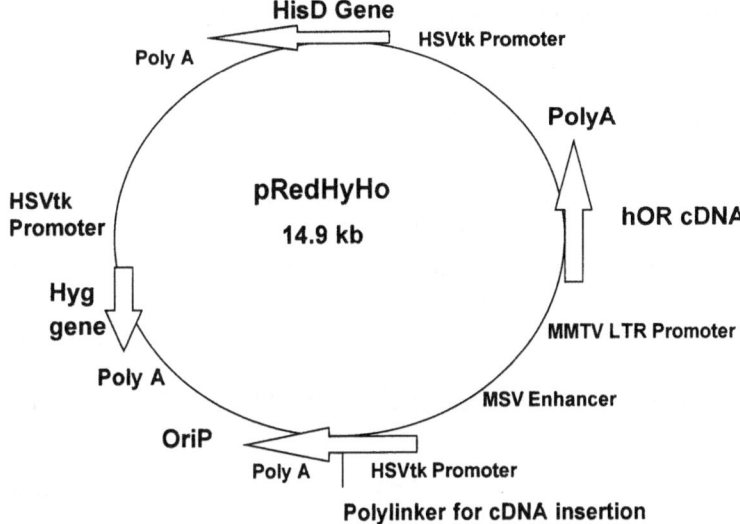

Fig. 2. Schematic map of the pRedHyHo vector. This vector contains five functional domains in addition to the sequences for replication and selection in bacteria. The vector was based originally on the pHEBo vector (Sugden et al. 1985). An expression unit containing the *Escherichia coli* HisD gene which confers resistance to l-histidinol, an expression unit for the P450 cDNA of interest and an expression unit for the human cytochrome P450 reductase cDNA (*hOR*) were introduced. The P450 cDNA, hygromycin B resistance; and the HisD gene all use the constitutive HSVtk promoter and polyadenylation signal. The promoter for hOR is the gluccocorticoid inducible MMTV LTR

and cytochrome P450 reductase was coexpressed. Our efforts at optimizing the expression system have resulted in our current, best cell line, designated h3A4/OR, containing an activity of 1000 pmol mg^{-1} min^{-1}, which is comparable to the average activity of a panel human liver microsomes (Table 1).

To speed future cell line development we have combined the elements of high and low copy number vector selection and cytochrome P450 reductase coexpression in to a single vector. This flexible expression vector has been designated pRedHyHo (schematic map contained in Fig. 2). This vector confers resistance to both hygromycin B

Table 1. Expression levels with human lymphoblasts – Comparison to human liver microsome

cDNA	Cell line designation	P450 content	Substrate	Lympho-blast activity	Human liver activity	Reference
CYP1A1	h1A1v2	25	7-Ethoxy-resorufin	155 ± 25	Non-hepatic	
CYP1A2	h1A2v2	40	7-Ethoxy-resorufin	73 ± 6	36 ± 20 74 ± 69	Murray et al. 1993 Yamazaki et al. 1993
CYP2A6	h2A3	55	Coumarin	780 ± 100	84 ± 36 15 ± 17	Murray et al. 1993 Yamazaki et al. 1993
CYP2B6	h2B6	60	7-Ethoxy-4-trifluoro-methyl-coumarin	190 ± 30	Not a specific substrate	
CYP2D6	h2D6v2	160	Bufuralol	935 ± 90	200 ± 80	Dayer et al. 1987
CYP2E1	h2E1/OR	40	p-Nitro-phenol	780 ± 40	490 ± 220	Murray et al. 1993
CYP3A4	h3A4/OR	20	Testosterone	1000 ± 170	1070 ± 1330	Yamazaki et al. 1993

Table contains cDNA designation, cell line designation for the human lymphoblastoid cell expressing the cDNA, microsomal P450 content (determined spectrophotometrically) in pmole P450 per mg microsomal protein, substrate used for comparison to human liver, activity with this substrate in human lymphoblast microsomes, in pmole per (mg × min), expressed as the mean and standard deviation for the last 6 batches (except CYP2E1 and CYP3A4 where data for only three batches are available), activity with this substrate for a human liver panel as the mean and standard deviation. Murray et al., (1993) contained data based on five liver specimens. Yamazaki et al. (1993) contained data based on 18 liver specimens. Dayer et al. (1987) is data for ten specimens from extensive metabolizers of debrisoquine

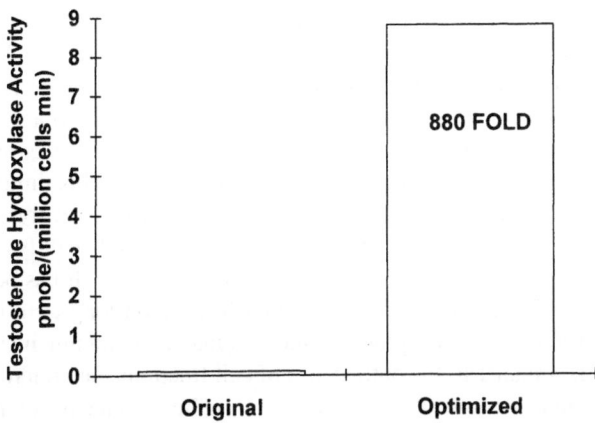

LOW COPY NUMBER VECTOR COMPARISON

Fig. 3. Comparison of the cellular testosterone 6β-hydroxylase activity in cells bearing pMF6 with CYP3A4 cDNA (*original*) and cells bearing pRedHyHo with CYP3A4 cDNA (*optimized*). pMF6 confers resistance to hygromycin B only; pRedHyHo expression was analyzed under hygromycin B selection. Activity is expressed as pmoles per million cells per minute. Approximately 7 million cells contain 1 mg total protein

and l-histidinol so that high and low copy numbers can be analyzed from a single transfected population. In addition, this vector contains a human cytochrome P450 OR cDNA which is expressed using a mouse mammary tumor virus long-term repeat promoter (MMTV-LTR) which is regulated by glucocorticoids. The strong murine sarcoma virus enhancer (MSV) influences expression of both the OR cDNA and the cytochrome P450 cDNA.

Figure 3 provides a comparison of CYP3A4 expression levels using the original vector adapted from pHEBo (pMF6) and the expression level for the pRedHyHo vector. For these comparisons, hygromycin B selection was used for both vectors. As Figure 3 demonstrates, a nearly 1000-fold improvement in catalytic activity has been achieved. CYP3A4 expression in pRedHyHo under l-histidinol selection has not yet been analyzed and hence is not contained in Fig. 1. Extrapolation based on vector copy number indicate that expression levels should be two- to threefold higher than the best cell line (h3A4/OR) contained in Fig. 1.

7.3.2 Development of Cell Fractionation Procedures

For many applications to drug metabolism, either whole cells, cell lysates, or subcellular fractions, can be used as active metabolizing elements. Each approach has its advantages and disadvantages. The use of whole cells allows the longest incubation times, and cofactors need not be added in order to observe metabolism. However, lower amounts of enzyme activity per unit volume is often observed. This limitation is likely due to nutrient depletion and limitation in oxygen transfer. In addition, cells must be prepared fresh each time. Whole cell lysates are simple to prepare and can be stored for subsequent use. It has been our experience that protein aggregates often form upon long-term incubation, particularly with agitation, introducing a maximum incubation time of less than 1 h. In contrast, use of cell fractions, i.e., microsomes, allows longer term incubations (up to 3 h) at higher enzyme concentrations (relative to both lysates and whole cells), and these preparations may be stored longer term and thawed for individual use providing a convenience to the user.

We found that standard microsome preparation procedures as used for human or rodent liver were unsuitable for isolating active enzymes from human lymphoblasts. Specific activities in microsomes were lower than for whole cell lysates. This loss of activity appeared to occur in other mammalian cell systems, which has led to the common use of whole cell lysates.

To identify where the activity was being lost we tracked the catalytic activity of the cytochrome P450 through the cell fractionation procedure and discovered that the majority of the catalytic activity was lost when the microsomes were pelleted during the 1-h high-speed spin. A simple modification to the procedure shortening the length of the centrifugation pelleting the microsomes has resulted in preparations retaining high specific activity (Penman et al. 1993). The shortening of the centrifugation time resulted in only a modest reduction in total protein yield.

In contrast to the instability of the expressed P450 protein when present as a microsomal pellet, once the microsomes are resuspended, they are stable for several years when stored at −80°C and they produce metabolites approximately linearly for 3 h. Figure 4 provides an example for p-nitrophenol hydroxylase with CYP2A6 microsomes. Similar stabilites have been observed for CYP2D6 (Penman et al. 1993),

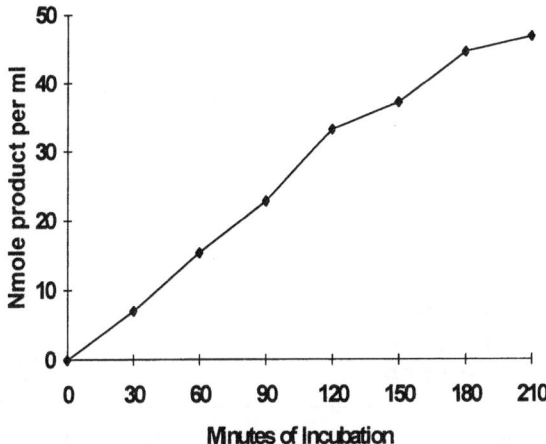

Fig. 4. Time course of *p*-nitrophenol hydroxylase activity by CYP2A6 microsomes. Assay conditions were 0.5 mg/ml protein, 1 m*M p*-nitrophenol, 1 mg/ml each of NADP$^+$ and glucose 6-phosphate and 0.4 U/ml glucose 6-phosphate dehydrogenase in 0.1 *M* Tris pH 7.5

CYP1A1, CYP1A2, CYP2B6, and CYP2E1 (unpublished data). CYP3A4 is somewhat less stable, with linear metabolite production for 1–1.5 h. The stability of the time dependence of metabolite production from the microsomes increases the versatility of the system.

7.3.3 Development cDNA-Expressing Cell Panel

Our goal has been to develop a comprehensive panel of human lymphoblastoid cell expressing cDNAs encoding individual, xenobiotic-metabolizing enzymes. Expression of the cDNAs at approximately 50 pmol P450/mg microsomal protein is desirable for maximum flexibility in applications cDNAs encoding human CYP1A1, CYP1A2, CYP2A6, CYP2B6, CYP2C8, CYP2C9, CYP2D6, CYP2E1, CYP3A4, and CYP3A5 have been transfected into the lymphoblastoid cells using the extrachromosomal vector system. Our typical approach has been to analyze expression levels at both high and low vector copy numbers and

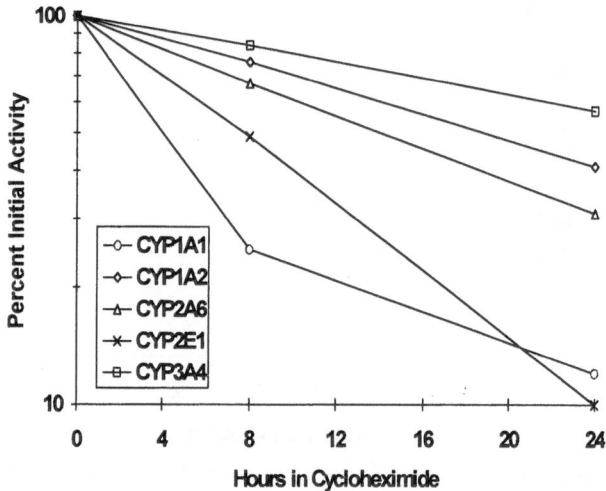

Fig. 5. Stability of cDNA expressed cytochrome P450 proteins in AHH-1 TK human lymphoblasts. Identity of the cDNAs is contained in the legend. New protein synthesis was blocked by the addition of 5 µg/ml cycloheximide. The loss of CYP-catalyzed activity was monitored by enzyme assay using coumarin 7-hydroxylase for CYP2A6, testosterone 6β-hydroxylase for CYP3A4, and 7-ethoxy-4-trifluoromethylcoumarin O-deethylase for CYP1A1, CYP1A2, and CYP2E1. Data are plotted relative to the activity without addition of protein synthesis inhibitor

then to proceed to further optimize expression levels via the use of multiple expression cassettes or OR coexpression.

We have observed a nearly 100-fold range in expression levels for different P450s using vectors which are otherwise identical. Differences in final expression level can be due to differences in transcription rate, mRNA stability, translation rate or protein stability. Figure 5 illustrates that different human cytochrome P450 proteins have different stabilities in the human lymphoblastoid cells. CYP1A1 and CYP2E1 are degraded relatively quickly while CYP2A6, CYP1A2, and CYP3A4 are substantially more stable.

Our observations suggest that sequences with in the coding regions of the cDNA can affect expression levels. Vector constructs containing the CYP2D6 cDNA show DMSO inducibility of catalytic activity. Isogenic constructs with other cDNAs do not show DMSO inducibility (except CYP2E1 where the effect is likely due to protein stabilization). DMSO induction of CYP2D6 expression likely reflects the presence of a DMSO responsive element in the cDNA.

We have taken a general approach of increasing cDNA expression through the use of high copy number vectors with multiple cDNA-containing expression units per vector. This approach works regardless of the specific mechanism behind the differences in expression levels. The vectors have the capacity for six independent expression units per vector before reaching a size which does not replicate well in bacteria. The largest vector actually used had four CYP3A4 expression units and resulted in a stable cell line. In practice, the limiting factor for expression level is cellular tolerance of high level cDNA expression. For example, some cDNAs, when highly expressed are lethal to the cells (for example: no viable cells were obtained when cells were transfected with a high copy number vector which constitutively expressed OR). Therefore one must determine, by producing multiple constructs, how much expression can be tolerated. Many of the cell lines expressing high levels of cDNAs (CYP1A2, CYP3A4 or CYP2E1) grow with a slower growth rate that control cells with vector only.

The mechanism for this apparent adverse effect on growth rate is at present unknown. However, cytochromes P450 can generate activated oxygen species which are known to be toxic to cells in culture. HepG2 cells expressing CYP2E1 have been developed to study this effect (Dai et al. 1993). We have observed similar effect in h2E1/OR cells which constitutively express CYP2E1 cDNA and express human cytochrome P450 reductase cDNA under control of a dexamethasone-inducible promoter. Exposure of the cells to dexamethasone induces a concentration-dependent increase in cell doubling time, an indicator of cytotoxicity (Fig. 6).

Considerable effort over the past 7 years has been directed towards reducing the magnitude of the differences in expression among the cDNAs such that expression levels are closer to approximately equal and that all cDNAs are expressed at or near the desired 50 pmol/mg microsomal protein level.

Our approach has most commonly utilized a sensitive catalytic assay for the P450 as a tool to assess the degree of expression with different

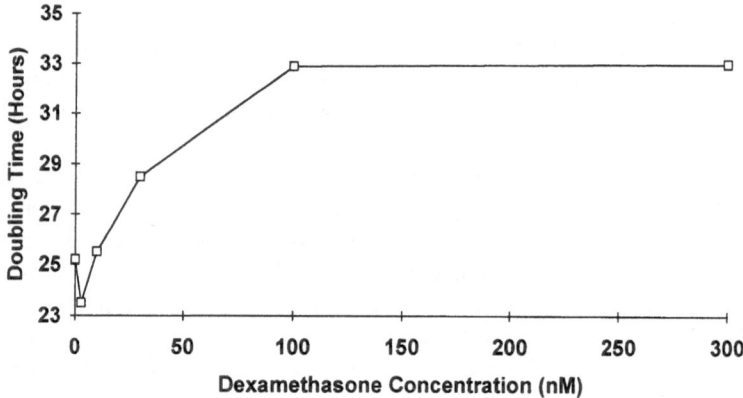

Fig. 6. Effect of dexamethasone treatment on cell doubling time for h2E1/OR cells which constitutively express CYP2E1 cDNA and express cytochrome P450 reductase with a promoter inducible by glucocorticoids. No difference in growth rate is observed for the first 3 days of growth in dexamethasone. During the next 3 days a pronounced, concentration-dependent increase in cell doubling time as a function of dexamethasone concentration was observed

vector constructs in bulk transfected populations, to screen clonal populations for activity and to asses stability of the cell line. Since the vector does not integrate, expression level is determined primarily by vector copy number and expression level per vector (not integration site). If all has gone well, minimal variation should be observed among clonal isolates (Crespi et al. 1993). Cytochrome P450 expression levels are then confirmed via western blot or spectra. Based on these results, the expression vector may be modified. For example, to introduce more expression units, the cells are retransfected and again analyzed.

Given the number of human P450 forms, this has been a substantial effort. Figure 7 summarizes the number of vector constructs which we have produced for each of ten different P450 cDNAs. The relatively high number of vector constructs for CYP3A4 reflects the importance of this enzyme in drug metabolism and hence, the desirability of developing a cell line with as high a CYP3A4 level as possible, and also, the difficulty in achieving this goal. In contrast, only a few constructs were necessary with CYP1A2 to achieve high expression levels. High-level, stable CYP1A1 expression recently has been achieved (Penman et al.

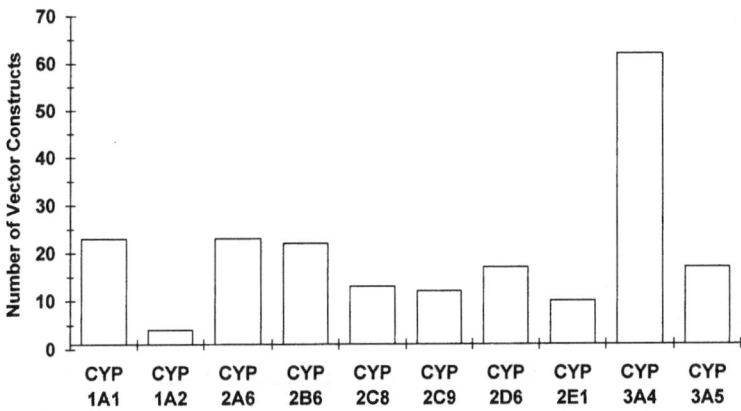

Fig. 7. Number of vector constructs produced to date for ten different human cytochrome P450 cDNAs

1994). For CYP2A6, high-level expression was achieved early (Crespi et al. 1990b), and the number vector constructs reflects our use of this cDNA, whose product is easily assayed, to evaluate new vector configurations. CYP2B6 has been expressed at high level, but expression, while stable for about 1.5 months, is not completely stable. The high number of vector constructs reflect our (to date) unsuccessful efforts to construct a more stable cell line with similar high levels of cDNA expression. CYP2C8, CYP2C9, and CYP3A5 are under development with promising preliminary results for both CYP2C9 and CYP3A5. CYP2D6 expressed quite well initially (Crespi et al. 1991b), and the vector constructs primarily reflect different vectors with multiple expression cassettes. The vector which worked best was utilized to develop h2D6v2 cells (Penman et al. 1993).

Table 1 and Fig. 8 summarize the P450 contents for cell lines expressing seven different CYP cDNAs. The cDNA which is expressed at the highest level is CYP2D6 (160 pmol/mg) cells. Cells containing CYP1A1, CYP1A2, CYP2D6, and CYP2E1 cDNAs contain two cDNA expression units/vector. CYP2E1 also contains OR coexpression.

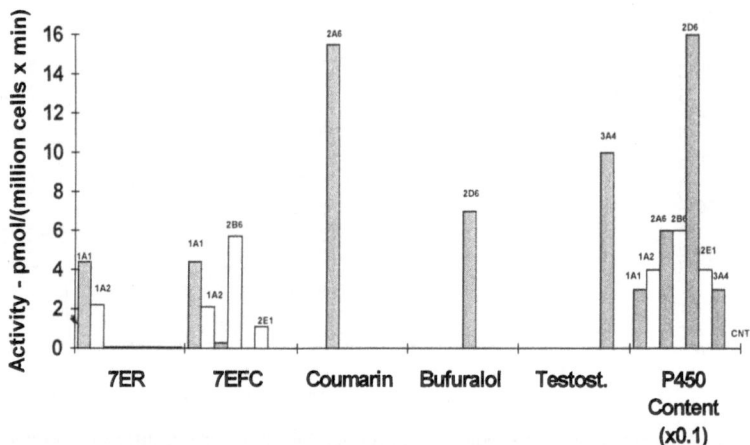

Fig. 8. Plot of catalytic activities for h1A1v2 (CYP1A1), h1A2v2 (CYP1A2), h2A3 (CYP2A6), h2B6 (CYP2B6), h2D6v2 (CYP2D6), h2E1/OR (CYP2E1), h3A4v3 (CYP3A4), and cHol (vector only control) cells. Bars alternate solid and white with the identity of nonzero values indicated at the top of the bar. P450 content is expressed as pmole per milligram of microsomal protein (× 0.1). *7ER*, 7-ethoxyresorufin O-deethylase; *7EFC*, 7-ethoxy-4-trifluoro-methylcoumarin O-deethylase

7.3.4 Current Status of the CYP cDNA Panel of Cell Lines

Table 1 contains microsomal P450 levels and catalytic activities for microsomes prepared from human cytochrome P450 cDNA expressing cell lines. This table also contains expression levels for human liver microsomes. The variation among independent human lymphoblast preparations is much lower than with human liver samples as should be expected when comparing a stable cell line to specimens from an out-bred population. Comparisons for the individual P450 forms are discussed below.

CYP1A1. Comparison of CYP1A1 to human liver levels is not appropriate as it is primarily an extrahepatic enzyme. Human CYP1A1 has been expressed in other mammalian cell systems, V79 cells, and human fibroblasts. The 7-ethoxyresorufin O-deethylase activity in V79

cells expressing CYP1A1 is 50 pmol mg^{-1} total protein min^{-1} (Schmalix et al. 1993), substantially lower than the 155 pmol mg^{-1} microsomal protein min^{-1} in lymphoblasts but roughly comparable to the lymphoblasts if total lymphoblast cell protein is assayed. The 7-ethoxyresorufin O-deethylase activity in human fibroblasts expressing CYP1A1 cDNA is 1.2 pmol 10^{-6} cells min^{-1} (States et al. 1993), about three- to fourfold lower than in h1A1v2 cells.

CYP1A2. The mean 7-ethoxyresorufin deethylase activity in microsomes from h1A2v2 cells (73 pmol mg^{-1} min^{-1} is comparable to that observed in human liver microsomes 36 pmol mg^{-1} min^{-1} (Murray et al. 1993) and 74 pmol mg^{-1} min^{-1} (Yamazaki et al. 1993). Human CYP1A2 has been expressed in other mammalian cell lines. The level of microsomal CYP1A2-catalyzed 7-ethoxyresorufin O-deethylase activity in h1A2v2 cells is about tenfold higher than in cytosol-free membranes prepared from V79 cells expressing the same cDNA (Wofel et al. 1992). The percentage recovery of protein in the V79 cell preparation procedure is probably higher than in our microsome preparation procedure but not sufficiently higher to completely compensate for the difference.

CYP2A6. The mean coumarin 7-hydroxylase activity in microsomes from h2A3 cells (780 pmol mg^{-1} min^{-1} is substantially above that observed in human liver microsomes 84 pmol mg^{-1} min^{-1}) (Murray et al. 1993) and 15 pmol mg^{-1} min^{-1} (Yamazaki et al. 1993). However, the low activity in human liver microsomes may be due to conducting the assay in phosphate buffers which can inhibit CYP2A6 (Crespi et al. 1990b; Pearce et al. 1992).

CYP2B6. The mean 7-ethoxy-4-trifluoromethylcoumarin deethylase activity in microsomes from h2B6 cells (190 pmol mg^{-1} min^{-1}). This substrate is not sufficiently specific to be used as a basis for comparisons of CYP2B6 levels in human liver microsomes (see Fig. 8). At the current time there is no specific assay for CYP2B6. CYP2B6 expression is highly variable among human liver samples (Yamano et al. 1989). Comparison based on western blot to only three human liver samples indicates that CYP2B6 levels in h2B6 cells are similar to those in human liver (Chang et al. 1993).

CYP2C. We have only begun to characterize the catalytic activity of CYP2C cDNA-expressing cells.

CYP2D6. The mean (+)-bufuralol 1'-hydroxylase activity in microsomes from h2D6v2 cells (935 pmol mg^{-1} min^{-1}) is about five times that observed in human liver microsomes (200 pmol mg^{-1} min^{-1}; Dayer et al. 1987). It is interesting to note that CYP2D6 cDNA expression is regulated by addition of DMSO to the media in human lymphoblasts (Crespi et al. 1991b; Penman et al. 1993).

CYP2E1. The mean *p*-nitrophenol hydroxylase activity in microsomes from h2E1/OR cells (cultured in the presence of dexamethasone to induce cytochrome P450 reductase activity; 780 pmol mg^{-1} min^{-1}) is nearly twice that observed in human liver microsomes (490 pmol mg^{-1} min^{-1}; Murray et al. 1993) and 12 times that observed for HepG2 cells expressing CYP2E1 cDNA (Dai et al. 1993). Note that both CYP2A6 and CYP2E1 catalyze *p*-nitrophenol hydroxylase; therefore human liver microsome *p*-nitrophenol hydroxylase values may be over estimated due to a contribution by CYP2A6.

CYP3A4. The mean testosterone 6β-hydroxylase activity in microsomes from h3A4/OR cells (cultured in the presence of dexamethasone to induce cytochrome P450 reductase activity; 1000 pmol mg^{-1} min^{-1}) is comparable to that observed in human liver microsomes (1070 pmol mg^{-1} min^{-1}; Yamazaki et al. 1993)

7.3.5 Applications to P450 Form-Specific Metabolism

One application of this system to drug development is the investigation of which cytochrome P450 form(s) is primarily responsible for the metabolism of a drug or drug candidate. Such information is used to establish whether a new drug is metabolized by an enzyme known to be polymorphic in human (i.e., CYP2D6) or is metabolized primarily by an enzyme known to be often involved in drug-drug interactions (i.e., CYP3A4 and to a lesser extent CYP2C9). Early information regarding cytochrome P450 form-specific metabolism can help to plan clinical investigations. For example, will it be important to phenotype or genotype individuals with respect to CYP2D6? Or, what are the likely drugs which may cause drug-drug interactions? (i.e., ketoconazole or rifampin for CYP3A4 and sulfaphenazole for CYP2C9).

The data in Fig. 6 provide examples of the specificity of individual P450 enzymes. Coumarin is hydroxylated in the 7 position exclusively by CYP2A6. Bufuralol is hydroxylated in the 1' position exclusively by CYP2D6. Testosterone is hydroxylated in the 6β position exclusively by CYP3A4. 7-Ethoxyresorufin is deethylated by both CYP1A enzymes. 7-Ethoxy-4-trifluoromethylcoumarin is deethylated by multiple cytochrome P450s including CYP1A1, CYP1A2, CYP2B6, CYP2A6, and CYP2E1.

Knowledge of the P450 contents of the cell lines allows correction for the differences in content among the cell lines. The data in Fig. 6 were obtained under saturating substrate concentrations. The relative affinities of the different enzymes is often an important parameter. For example, if CYP3A4 is a low-affinity enzyme, drug-drug interactions with CYP3A4 are not significant. Therefore it is often desirable to test multiple drug concentrations in order to assess the relative affinities of the different enzymes.

A second application is testing interaction with specific enzymes by competitive inhibition analyzes. Again, CYP2D6 and CYP3A4 are enzymes of particular interest for this approach. cDNA-expressed enzymes offer an advantage over liver microsomes because of their consistency among preparations and the lack of potentially competing enzymes. We have analyzed inhibition of CYP2D6-catalyzed bufuralol 1'-hydroxylase and CYP3A4-catalyzed testosterone 6β-hydroxylase activities. Table 2 contains apparent K_i data for model compounds with these two enzymes. Known inhibitors of CYP2D6, quinidine, perhexiline, dl-propranalol, dextromethorphan, and sparteine, inhibited cDNA-expressed CYP2D6 enzyme. Known inhibitors of CYP3A4, ketoconazole, (+)-miconazole, terfenadine, and erythromycin, inhibited cDNA-expressed CYP3A4. The relative potency of the inhibition is in general agreement with that observed in other systems.

Table 2. Competitive inhibitors of CYP2D6 and CYP3A4

Enzyme	Compound	Apparent K_i (μM)
CYP2D6	Quinidine	0.05
CYP2D6	Perhexiline	0.12
CYP2D6	*dl*-Propranalol	1.1
CYP2D6	Dextromethorphan	21
CYP2D6	Sparteine	80
CYP3A4	Ketoconazole	0.02
CYP3A4	(+)-Miconazole	0.8
CYP3A4	Terfenadine	3.4
CYP3A4	Erythromycin	70

Inhibition of CYP2D6 was measured using bufuralol 1'-hydroxylase. Specific assay conditions used a 10 minute incubation with 0.02 mg microsomal protein in 0.2 ml with an NADPH generating system. Bufuralol concentrations were 0.01, 0.03 and 0.1 m*M*. Inhibition of CYP3A4 was measured using testosterone 6β-hydroxylase. Specific assay conditions used a 45 minute incubation with 0.36 mg microsomal protein in 0.5 ml with an NADPH generating system. Testosterone concentrations were 0.04, 0.12 and 0.36 m*M*

7.4 Future Directions

The field of xenobiotic metabolism is rich, with a myriad of different and often competing pathways of metabolism. Our efforts at cDNA expression have only begun to develop a comprehensive system. Several areas are under active development in our laboratory. These include:

1. CYP2C expression. The CYP2C subfamily has not been developed in the lymphoblast expression system. Considerable progress has recently been made with CYP2C9.
2. CYP3A expression. Cell lines with higher CYP3A4 expression are desirable because of the central role of this enzyme in drug metabolism. In addition, other CYP3A enzymes (i.e., CYP3A5) have not been fully developed in the system.

3. Phase II enzymes are often as important as cytochrome P450 enzymes in the metabolism of drug and other xenobiotics. Except for microsomal epoxide hydrolase, these enzymes remain to be developed in the lymphoblast system.
4. Coexpression of multiple enzymes. The human lymphoblast cell/extrachromosomal vector system is uniquely suited for the coexpression of multiple enzymes (Crespi et al. 1991c). Coexpression of five procarcinogen-metabolizing enzymes has been achieved. It will be desirable to coexpress, at high levels, multiple drug-metabolizing enzymes. This will allow metabolite profiling using a single, stable cell line.

References

Burke MD, Prough RA, Mayer RT (1977) Characteristics of a microsomal cytochrome P-448-mediated reaction Ethoxyresorufin O-de-ethylation. Drug Metab Disp 5:1–8

Chang TKH, GF Weber, CL Crespi and DJ Waxman (1993) Differential activation of cyclophosphamide and ifosphamide by cytochromes P-450 2B and 3A in human liver microsomes. Cancer Res 53:5629–5637

Crespi CL (1991) Expression of cytochrome P450 cDNAs in human B lymphoblastoid cells for applications to toxicology and metabolite analysis. In: Waterman MR, Johnson EF (eds) Methods in enzymology, vol 206. Academic, New York, pp 123–129

Crespi CL (1992) Method and a preparation to select for transfected DNA in mammalian cells. US patent no 5,128,255

Crespi CL, Thilly WG (1984) Assay for mutation in a human lymphoblastoid line, AHH-1, competent for xenobiotic metabolism. Mutat Res 128:221–230

Crespi CL, Altman JD, Marletta MA (1985) Xenobiotic metabolism in a human lymphoblastoid cell line. Chem Biol Interact 53:257–272

Crespi CL, Steimel DT, Aoyoma T, Gelboin HV, Gonzalez FJ (1990a) Stable expression of human cytochrome P450IA2 cDNA in a human lymphoblastoid cell line: role of the enzyme in the metabolic activation of aflatoxin B1Mol. Carcinogenesis 3:5–8

Crespi CL, Penman BW, Leakey JAE, Arlotto MP, Start A, Turner T, Steimel D, Rudo K, Davies RL, Langenbach R (1990b) Human cytochrome P450IIA3: cDNA sequence, role of the enzyme in the metabolic activation of promutagens, comparison to nitrosamine activation by human cytochrome P450IIE1. Carcinogenesis 8:1293–1300

Crespi CL, Langenbach R, Penman BW(1990c) The development of a panel of human cell lines expressing specific human cytochrome P450 cDNAs. In: Mendelsohn ML, Albertini RL (eds) Mutation and the environment, vol 340D. Liss, New York, pp 97–106

Crespi CL, Penman BW, Steimel DT, Gelboin HV, Gonzalez FJ (1991a) The development of a human cell line stably expressing human CYP3A4: role in the metabolic activation of aflatoxin B1 and comparison to CYP1A2 and CYP2A3. Carcinogenesis 12:355–359

Crespi CL, Penman BW, Gelboin HV, Gonzalez FJ (1991b) A tobacco smoke-derived nitrosamine, 4-(methylnitrosamino)-1-(3-pyridyl)-1-butanone, is activated by multiple human cytochrome P450s including the polymorphic human cytochrome P4502D6. Carcinogenesis 12:1197–1201

Crespi CL, Gonzalez FJ, Steimel DT, Turner TR, Gelboin HV, Penman BW, Langenbach R (1991c) A metabolically competent cell line expressing five cDNAs encoding procarcinogen-activating enzymes: application to mutagenicity testing. Chem Res Toxicol 4:566–572

Crespi CL, Langenbach R, Penman BW(1993) Human cell lines, derived from AHH-1 TK human lymphoblasts, genetically engineered for expression of cytochromes P450. Toxicology 82:89–104

Dai Y, Rashba-Step J, Cederbaum AI (1993) Stable expression of human cytochrome P4502E1 in HepG2 cells: characterization of catalytitic activities and production of reactive oxygen intermediates. Biochemistry 32:6928-6937

Davies RL, Crespi CL, Rudo K, Turner TR, Langenbach R (1989) Development of a human cell line by selection and drug-metabolizing gene transfection with increased capacity to activate procarcinogens. Carcinogenesis 10:885–891

Dayer P, Kronbach T, Eichelbaum M, Meyer UA(1987) Enzymatic basis of the debrisoquine/sparteine-type polymorphism of drug oxidation. Biochem Pharmacol 36:4145–4152

DeLuca JG, Dysart GR, Rasnick D, Bradley MO (1988) A direct, highly sensitive assay for cytochrome P-450 catalyzed O-deethylation using a novel coumarin analog. Biochem Pharmacol 39:1731–1739

Gonzalez FJ (1988) The molecular biology of cytochrome P450s. Pharmacol Rev 40:243–288

Gonzalez FJ, Crespi CL, Gelboin HV (1991) cDNA-expressed human cytochrome P450s: a new age of molecular toxicology and human risk assessment. Mutat Res 247:113–127

Greenlee WF, Poland A (1978) An improved assay of 7-ethoxycoumarin O-deethylase activity: induction of hepatic enzyme activity in C57BL/6J and

DBA/2J mice by phenobarbital, 3-methylcholanthrene and 2,3,7,8-tetrachlorodibenzo-p-dioxin. J Pharmacol Exp Ther 205:596–605

Guengerich FP (1988) Roles of cytochrome P-450 enzymes in chemical carcinogenesis and cancer chemotherapy. Cancer Res 48:2946–2954

Komori M, Kitamura R, Fukuta H, Inoue H, Baba H, Yoshikawa K, Kamataki T (1993) Transgenic Drosophila carrying mammalian cytochrome P-4501A1: an application to toxicology testing. Carcinogenesis 14:1683–1688

Kronbach T, Mathys D, Gut J, Catin T, Meyer UA (1987) High-performance liquid chromatographic assays for bufuralol 1'-hydroxylase, debrisoquine 4-hydroxylase and dextromethorphan O-demethylase in microsomes and purified cytochrome P450 isozymes of human liver. Anal Biochem 162:24-32

Langenbach RL, Smith PB, Crespi C (1992) Recombinant DNA approaches for the development of metabolic systems used in in vitro toxicology. Mutat Res 277:251–275

Lowry OH, Rosebrough NJ, Farr AL, Randall FJ (1951) Protein measurement with Folin phenol reagent. J Biol Chem 62:315–323

Murray BP, Edwards RJ, Murray S, Singleton AM, Davies DS, Boobis AR (1993) Human hepatic CYP1A1 and CYP1A2 content, determined with specific anti-peptide antibodies, correlates with the mutagenic activation of PhIP. Carcinogenesis 14:585–592

Omura T, Sato R (1964) The carbon monoxide-binding pigment of liver microsomes. II. Solubilization, purification and properties. J Biol Chem 239:2379–2385

Pearce R, Greenway D, Parkinson A (1992) Species differences and interindiviual variation in liver microsomal cytochrome P450 2A enzymes: effects on coumarin, dicoumarol and testosterone oxidation. Arch Biochem Biophys 298:211–255

Penman BW, Crespi CL (1987) Analysis of human lymphoblast mutation assays by using historical negative control data bases. Environ Mol Mutagen 10:35–60

Penman BW, Reece J, Smith T, Yang CS, Gelboin HV, Gonzalez FJ, Crespi CL (1993) Characterization of a human cell line expressing high levels of cDNA-derived CYP2D6. Pharmacogenetics 3:28–39

Penman BW, Chen L, Gelboin HV, Gonzalez FJ, Crespi CL (1994) Development of a human lymphoblastoid cell line constitutively expressing human CYP1A1 cDNA: substrate specificity with model substrates and promutagens. Carcinogenesis (submitted)

Schmalix WA, Maser H, Kiefer F, Reen R, Wiebel FJ, Gonzalez F, Seidel A, Glatt H, Greim H, Doehmer J (1993) Stable expression of human cytochrome P450 1A1 cDNA in V79 Chinese hamster cells and metabolic activation of benzo(a)pyrene. Eur J Pharmacol 248:251–261

States JC, Quan T, Hines RN, Novak RF, Runge-Morris M (1993) Expression of human cytochrome P450 1A1 in DNA repair deficient and proficient human fibroblasts stably transformed with an inducible expression vector. Carcinogenesis 14:1643–1649

Sugden B, Marsh K, Yates J (1985) A vector that replicates as a plasmid and can be efficiently selected in B-lymphoblasts transformed by Epstein-Barr virus. Mol Cell Biol 5:410–413

Waterman MR, Johnson EF (1991) Cytochromes P450. Academic, New York (Methods in enzymology, vol 206)

Waxman, DJ, Ko A, Walsh C (1983) Regioselectivity and stereoselectivity of androgen hydroxylations catalyzed by cytochrome P-450 isozymes purified from phenobarbital-induced rat liver. J Biol Chem 258:11937–11947

Wolfel C, Heinrich-Hirsch B, Schulz-Schlage T, Seidel A, Frank H, Ramp U, Wachter F, Wiebel FJ, Gonzalez F, Greim H, Doehmer J (1992) Genetically engineered V79 Chinese hamster cells for stable expression of human cytochrome P4501A2. Eur J Pharmacol 228:95–102

Yamano S, Nhamburo PT, Aoyama T, Meyer UA, Inaba T, Kalow W, Gelboin HV, McBride OW, Gonzalez FJ (1989) cDNA cloning and sequence and cDNA directed expression of human P450 IIB1: identification of a normal and two variant cDNAs derived from the CYP2B locus on chromosome 19 and differential expression of the IIB mRNAs in human liver. Biochemistry 28:7340–7348

Yamazaki H, Mimura M, Oda Y, Inui Y, Shiraga T, Iwasaki K, Guengerich FP, Shimada T (1993) Roles of different forms of cytochrome P450 in the activation of the promutagen 6-aminochrysenes to genotoxic metabolites in human liver microsomes. Carcinogenesis 14:1271–1278

Yoakum GH (1984) Protoplast fusion: a method to transfect human cells for gene isolation, oncogene testing and construction of specialized cell lines. Biotechniques 1/2:24–30

8 Metabolic Reactions and Recombinant Isoenzymes of Cytochrome P450: Information Generated and Value for Pharmaceutical Development

R. E. Tynes

8.1   Introduction 135
8.2   Pharmaceuticals and P450: Interest and Value 137
8.2.1   Eleven Relevant P450s 137
8.2.2   Application Hierarchy 138
8.2.3   Information and Value Generated 139
8.3   Individual Isoenzymes: The Vital Statistics 142
8.3.1   Family I Isoenzymes 142
8.3.2   Family II Isoenzymes 145
8.3.3   Family III Isoenzymes.................................. 148
8.4   Molecular Biology: Recombinant Isoenzymes 150
8.5   Conclusions 154
References 155

8.1 Introduction

Cytochrome P450s are heme-containing mono-oxygenases which encompass a multi-isoenzyme superfamily whose chemical hydroxylation specificity includes xenobiotics as important as pharmaceuticals and carcinogens (Gonzalez 1990, 1992; Guengerich 1992; Nelson et al. 1993). Different isoenzyme members of this family generally fall into one of two functional classifications. Roughly 10–20 isoenzymes participate primarily in the metabolism of foreign chemicals, for example, drugs, while numerous other isoenzymes participate in specific endogenous functions of catabolic or anabolic nature. Drug-metabolizing isoenzymes are the focus of this report. Cytochrome P450 catalysis involves multiple proteins; usually assisting in electron transfer are cytochrome P450 reductase and occasionally cytochrome b_5 and cytochrome b_5 reductase. All are localized to the endoplasmic reticulum, which can be isolated after cell fractionation as microsomes. The liver has high P450 concentration, but P450s have near universal distribution, including but not limited to kidney, lung, and intestine, which are barrier organs to chemical entry.

Cytochrome P450s are major players in the so-called "phase I" of drug metabolism: the NADPH-dependent oxidation reactions that generate hydroxylated, dealkylated, or epoxidated drug. This first step in drug metabolism is critical for the major goal of biotransformation, which increases the water solubility of a drug in preparation for excretion and elimination. These initial hydroxylated metabolites are frequently carried down subsequent pathways of metabolism: the so-called "phase II" or conjugative reactions, which result in glucuronide-, sulfate- or glutathione-conjugated drug, often the end-product of excretion. Since it is the P450 that is situated first in this "catalytic cascade" of drug metabolism, sometimes the isoenzyme acts as a control point influencing drug plasma level, half-life, elimination, or final effect. For biopharmaceutics the relevance of P450 is readily apparent.

With 10–20 different isoenzymes participating in drug metabolism, an obvious complexity arises in drug study. Identifying individual isoenzyme reactions is nonetheless central for the evaluation of a drug candidate. The clearest example is when metabolism is carried out by an isoenzyme which is genetically polymorphic in man. This occurs for isoenzyme CYP2D6 and one isoenzyme of the CYP2C subfamily, which are responsible for the clinical debrisoquine/sparteine and mephenytoin polymorphisms, respectively (Meyer et al. 1990; Wrighton et al. 1993). Isoenzymes can also show large interindividual variability for nongenetic reasons. In any case predictive capacity is at hand when drug candidates are examined with isoenzymes early in the development scheme. With the target isoenzyme identified, a rational and cost-effective development scheme can be envisioned for clinical drug-drug interaction study and patient variability concerns. Isoenzyme application contributes as well to the determination of metabolite patterns and is ideal for reaction kinetics and in scale-up for the production of defined metabolites. It is increasingly appreciated that the best way to study individual isoenzymes is first to produce them through recombinant techniques. Recombinant cell lines have been developed that express human P450 isoenzymes. Their use by the pharmaceutical industry will ultimately translate into a qualitative improvement and streamlining of the drug design through development process.

8.2 Pharmaceuticals and P450: Interests and Value

8.2.1 Eleven Relevant P450s

Primary consideration can be given to the 11 most important human P450 isoenzymes whose function is generally thought to focus on xeno-

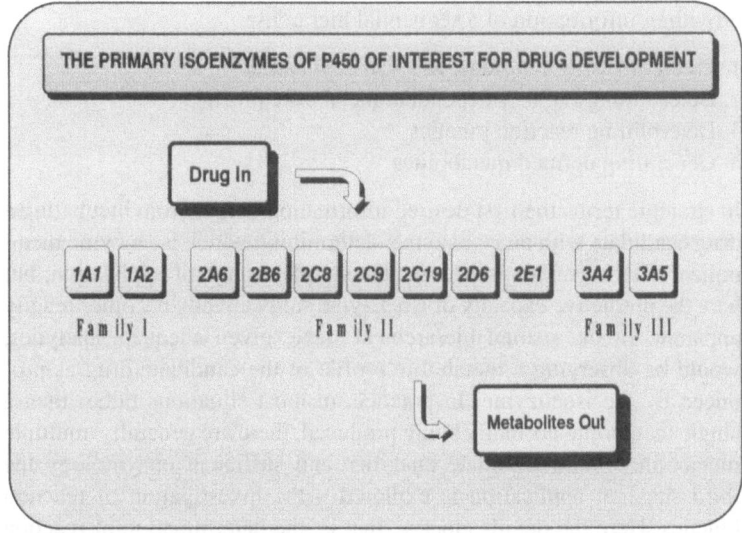

Fig. 1. The 11 major human cytochrome p450 isoenzymes implicated in the metabolism of pharmaceuticals

biotic metabolism (Fig. 1). Drug-metabolizing P450 isoenzymes are not restricted, either evolutionarily or structurally, to a single gene family but rather are spread over three gene families: the families I, II, and III (Nelson et al. 1993). The family I members (CYP1A1 and CYP1A2) are generally associated with procarcinogen activation of aryl and arylamine hydrocarbons, constituents from combustion processes. The gene family II (CYP2) is the largest in man. At least five subfamilies therein are relevant to drug metabolism – CYP2A, CYP2B, CYP2D, CYP2C, and CYP2E. Family III includes CYP3A4, likely the human drug hydroxylase of widest relevance due to a remarkably accommodating active site imparting diverse substrate acceptance. The interest in P450 for biopharmaceutics is clear; namely, the basic process of "drug in" through "metabolites out."

8.2.2 Application Hierarchy

For qualitatively improving drug development and facilitating drug design, the use of isoenzyme in pharmaceuticals can be envisioned as providing information of a sequential hierarchy:

1. Predicting drug candidate as P450-X substrate
2. Determining P450-X dependent metabolite profiles
3. Determining reaction kinetics
4. Generating defined metabolites

In strategic terms the first desired information comes from incubating a drug candidate with an isoenzyme: determining which isoenzyme metabolizes the drug. This is perhaps the simplest stage of application, but here the predictive capacity of isoenzyme study already becomes readily apparent. In the second hierarchical stage, given adequate analytics, would be observing a metabolite profile of the candidate drug as produced by the isoenzyme. In practice, in most situations rather than a single metabolite normally being produced, there are generally multiple metabolites. With adequate analytics and sufficient enzymology the third stage of application is explored – the investigation of reaction kinetics. Here the details emerge, that is, the determination of reaction affinity and reaction velocity. Again, the level of required results interpretation, namely the meaning of the information for drug biopharmaceutics, is elevated to the higher plateau. The last exploitation, hierarchy stage 4, could be the up-scale generation of the metabolites themselves. Information concerning structure and possible metabolite activity are generally already requirements for most registration documentation.

8.2.3 Information and Value Generated

Identifying the Isoenzyme. Having once identified a drug candidate as a substrate for P450 isoenzyme-X (hierarchy stage 1), what is the information generated, and in a more general sense, what is its value to drug study (Fig. 2). The isoenzyme identified, for example; will have a certain abundance in liver, with this abundance information increasingly available in quality from a variety of studies. Second, the isoenzyme will or will not have an extrahepatic distribution, again, information continually more available from the literature. Furthermore, with an increasing understanding of gene regulation, certain isoenzymes are characterized as having an inducibility, that is, they can be elevated upon exposure to an inducer. In addition, an expanding database of known drug substrates for individual isoenzymes is accumulating, and by which a new drug candidate could rapidly be coclassified. The implications of such information lie in its value. The value of isoenzyme identification is simplest in one case, the early recognition of a potential drug polymorphism. At least two human drug-metabolizing P450 isoenzymes are genetically polymorphic in the human population. When one of these two isoenzymes is identified as major for the metabolism of a drug, deficient individuals will thus lack this metabolism pathway, and consequences for drug efficacy, clinical trial patient selection, or toxicity throughout the population must be reconsidered. From the information of isoenzyme tissue location site(s) of metabolism can be proposed. Inducibility information provides a predictive capacity as to how the coadministration of known drug inducers would influence metabolism. Several human P450 isoenzymes are also highly variable in man, but for nongenetic reasons. From this information interindividual variability can be predicted, as relevant, but also as restricted, to the significance of that specific isoenzyme in the overall scheme of biopharmaceutics.

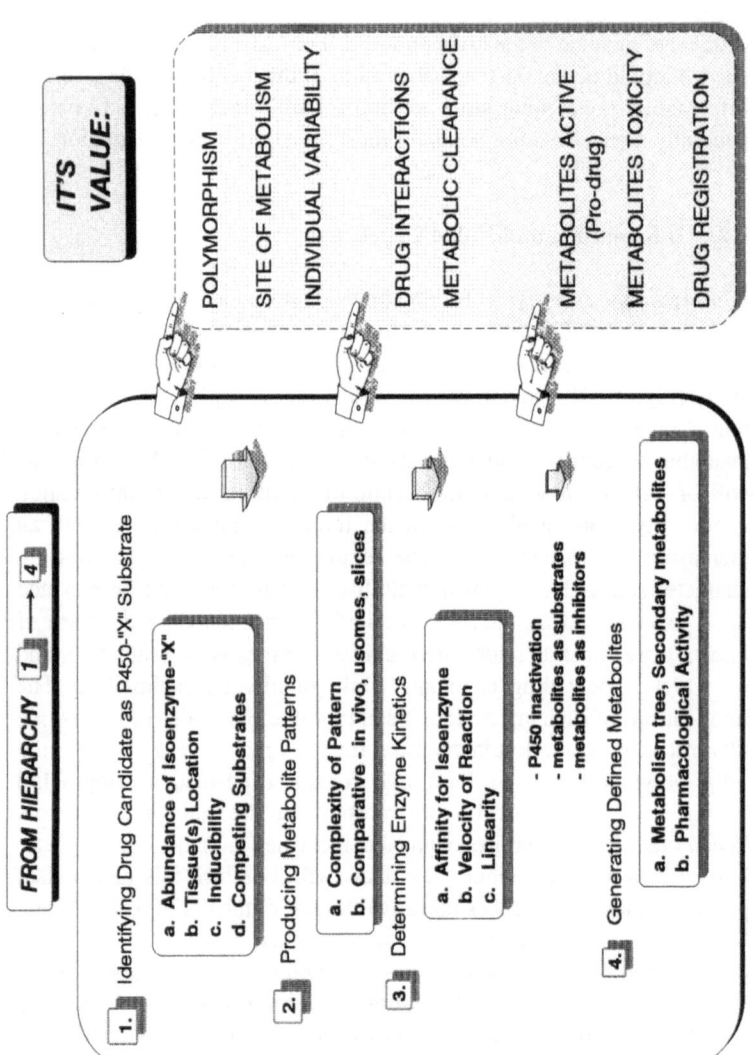

Fig. 2. The progressive hierarchy of cytochrome p450 isoenzyme use in drug metabolism

Metabolite Patterns. For producing metabolite patterns isoenzyme application furnishes information concerning the complexity of the pattern and can be used for comparative purposes with metabolite patterns obtained from other sources. A comparison with human liver microsomes would expose metabolites generated from multiple P450 isoenzymes whose formation could subsequently be modeled using single isoenzymes, either mixed or in sequential incubation. Hepatocyte and tissue slice comparison would reveal potential conjugative metabolites; namely, double-metabolism products of first phase I and then phase II reactions. The comparison of metabolite patterns from human isoenzymes and rodents can facilitate the rational selection of experimental models for safety assessment. In the ideal situation is the human isoenzyme metabolite pattern identical to that found in vivo in man. When this occurs (see below) the dependence of the in vivo situation on that isoenzyme is readily modeled.

Reaction Kinetics. The determination of reaction kinetics between isoenzyme and candidate drug provides information as to the efficiency of catalysis, as measured by K_m and V_{max}. Combined with isoenzyme abundance and drug level in vivo, the value of these parameters lies in defining one isoenzyme's significance when one drug reaction can be catalyzed by multiple isoenzymes. The affinity of the drug-isoenzyme reaction is also critical for predicting drug-drug interactions as they occur at the site of a target isoenzyme. In isoenzyme-drug candidate reactions sometimes the metabolites themselves serve as substrates or inhibitors. Metabolite-directed P450 inactivation also occasionally occurs, with obvious implications for drug interactions. Given the various known modes of P450 inhibition – competitive, noncompetitive, and enzyme inactivation – the isoenzyme study adds value for assessing the mechanism of inhibition and therefore the varying implications depending on the inhibition mode. Strategically the ultimate goal of reaction kinetics would be their direct insertion into physiologically based pharmacokinetic models after the critical steps of biopharmaceutics have been identified in animal models.

Defined Metabolites. After establishing which isoenzyme metabolizes which candidate drug, determining its metabolite patterns, and defining the reaction kinetics, in application the isoenzyme can additionally contribute to the generation of defined metabolites. The absolute use-fulness of an isoenzyme system for this application depends on the biocatalytic capacity of the isoenzyme expression method, the velocity of drug turnover, and the availability or nonavailability of competing synthetic methods. It is obvious that the evaluation of the drug metabo-lites themselves in the absence of the parent drug has value for deter-mining whether the metabolites are pharmacologically active or toxic. The concept of P450 prodrug activation has been illustrated for one example, the bioactivation of the narcotic codeine through CYP2D6-dependent O-demethylation (Dayer et al. 1988; Mikus et al. 1991). Structure elucidation of the major metabolites is usually a prerequisite for drug registration in any case. The production of analytical stand-ards is an additional potential value achievable with isoenzymes. Enzy-matic methods of metabolite synthesis can be advantageous where the metabolites are a complex synthesis, of moderate stability, or too numerous for routine synthesis. Further advantageous is that enzymes are typically regio-specific; moreover, the P450 isoenzymes are note-worthy for often accepting only a particular chirality as substrate and for forming a stereospecific metabolite. The use of isoenzyme bioreac-tors to produce drugs is also not unconceivable.

8.3 Individual Isoenzymes: The Vital Statistics

Having identified an individual isoenzyme that interacts with a candi-date drug, the interpretive value then has a basis on what is known about that isoenzyme. This includes possible gene polymorphisms, absolute liver levels, variabilities, regulation through induction, known sub-strates and inhibitors. This information, that is the isoenzymes vital statistics, is presented in outline (Fig. 3).

8.3.1 Family I Isoenzymes

8.3.1.1 Human Isoenzyme P450 1A1

The CYP1A1 is most affectionately known as aryl hydrocarbon hydrox-ylase. It was initially characterized as that isoenzyme most involved in the activation of procarcinogens of the polyclyclic hydrocarbon class.

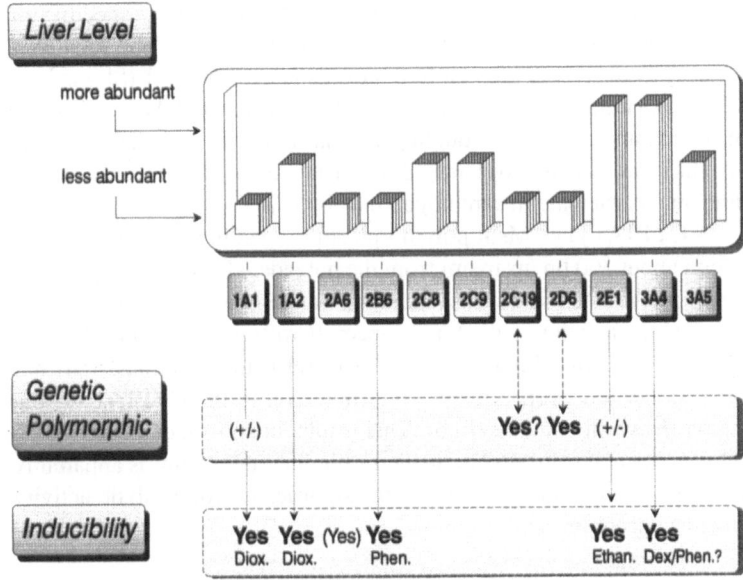

Fig. 3. Overview of main cytochrome p450 isoenzymes: liver levels, genetic polymorphisms, and inducibility in man

The gene regulation of CYP1A1 has continued to receive much attention as it is inducible by exposure to several polycyclic aromatic hydrocarbons, and by dioxin as well, one of the most toxic chemicals known in public health (Fujii-Kuriyama et al. 1992). Gene induction is controlled by a recently identified Ah receptor (Burbach et al. 1992). This receptor binds after heterodimerization with the ARNT protein to dioxin or so-called xenobiotic response elements upstream on the CYP1A1 promotor to up-regulate gene transcription. Drugs on the market are not known which induce CYP1A1 via an Ah-receptor mechanism, with the possible exception of omeprazole (Daujat et al. 1992). Naphthoflavone is a specific inhibitor of CYP1A1.

CYP1A1 is found only in trace amounts in the human liver. Polymerase chain reaction analysis has revealed the constitutive nature of CYP1A1 mRNA expression in a variety of adult and fetal cells and tissues, including freshly isolated lymphocytes and pulmonary macro-

phages (Omiecinski et al. 1990). A monoclonal antibody specific for
1A1 has identified CYP1A1 protein in both human adult and fetal liver
(Murray et al. 1992). Human lung also contains CYP1A1 (Shimada et
al. 1992). Human lung contains about 10 pmol total P450 per milligram
of microsomes, which is roughly 2% that found in liver. Of probably
insignificant role for systemic drug metabolism, CYP1A1 in human
lung may participate in carcinogen metabolism.

A CYP1A1 gene MspI polymorphism located in the 3' untranslated
region about 300 bp upstream from the polyadenylation signal has been
found in African-Americans; its significance is unknown and has no
apparent association with lung cancer (Crofts et al. 1993). Closely
linked to a second MspI polymorphism (also in 3' untranslated) has
been found, however, a novel point mutation in the CYP1A1 coding
region (Hayashi et al. 1991a,b). This results in an isoleucine for valine
substitution at residue 462 in the heme binding region; this is apparently
associated with a lung cancer, but the significance for catalytic activity
remains uncertain.

8.3.1.2 *Human Isoenzyme P450 1A2*

CYP1A2 is the principle phenacetin O-deethylase in man (Sesardic et
al. 1989). It is also responsible for the metabolism of promutagens
including heterocyclic amines and arylamines (Butler et al. 1989). The
caffeine breath test using [3-^{14}C]caffeine has been proposed in addition
to phenacetin as an in vivo probe for CYP1A2 activity in man (Lambert
et al. 1990). The breath test appears to show positive correlation with
cigarette smoking dose. CYP1A2 appears to be coinduced with
CYP1A1, primarily by an Ah receptor mechanism, explaining this
correlation. CYP1A2 is constitutively expressed in human liver. Interindividual variation of CYP1A2 level in liver has been reported at tenfold
based on immunoblotting (Sesardic et al. 1990). Phenacetin measurements in vivo have demonstrated a 30-fold interindividual variability
for that activity. Furafylline is a specific CYP1A2 inhibitor.

8.3.2 Family II Isoenzymes

8.3.2.1 Human Isoenzyme P450 2A6

This isoenzyme is responsible for almost all human liver coumarin 7-hydroxylase activity and apparently can also metabolize diethylnitrosamine (Pearce et al. 1992). There appear to be no major drugs on the market which are metabolized predominantly by CYP2A6 in vivo. Human CYP2A6 is a minor isoenzyme, accounting for a maximum of 1% of total hepatic cytochrome P450 (Yun et al. 1991). The variability of CYP2A6 expression in man is extremely large. This isoenzyme has shown a 100-fold variability among individuals based on immunoblotting measurements, with some livers having nearly undetectable levels. A 17-fold variability has been found in activity rates of microsomal coumarin 7-hydroxylase. In the mouse the 2A and 2B gene subfamilies are coinduced by phenobarbital. This coregulation also appears to occur in man; since livers with high level of one invariably also have relatively high levels of the other. CYP2A7 is a mutant allele of CYP2A6 also found in liver, but its significance is uncertain. There are no known selective CYP2A6 inhibitors suitable for use in vivo.

8.3.2.2 Human Isoenzyme P450 2B6

In the rat the P450 2B subfamily has had historical significance as "the phenobarbital-induced P450" (Waxman and Azaroff 1992). In man CYP2B6 also appears to be barbiturate inducible. CYP2B6 appears to be present at low level in liver, about 1% of total P450, and has been partially purified from human liver (Mayumi et al. 1993). Immunoblotting of 50 livers has revealed detectable CYP2B6 in only 12 samples; of which only four had relatively high levels, and these four also had high CYP3A4 content. In rats the 2B and 3A isoenzymes are coinduced by phenobarbital and dexamethasone, and CYP2B6 and CYP3A4 could as well be linked in P450 regulation. Experimentation on 13 liver samples showed the level CYP2B6 mRNA and protein to exhibit substantial variation. Immunoblotting measurements have described 2B6 level variation as large.

The human CYP2B6 mRNA appears to be inefficiently processed, leading to a high level of aberrant message but with enough functional mRNA to produce some enzyme (Yamano et al. 1989). Inefficient processing does not appear to lead to interindividual variability; the ratio of properly to improperly spliced product remains similar between livers and among different cell and tissue types. CYP2A6 has relatively low activity towards typical P450 substrates such as 7-ethoxycoumarin, ethoxy- and pentoxyresorufin, ethylmorphine, and benzphetamine (Mayumi et al. 1993). That CYP2B6 is a minor constituent of liver and has low activity suggests that 2B biotransformations in man may not be as significant as the popular phenobarbital-induced experimental animal models suggest.

8.3.2.3 Human Isoenzymes 2C8, 2C9, and 2C18/19

The CYP2C is the most complex subfamily in man. Within this subfamily is one isoenzyme responsible for the clinically significant mephenytoin polymorphism. This second genetic polymorphism was first noted from individuals deficient in 4'-hydroxylation of the S-enantiomer of the anticonvulsant mephenytoin. The major liver 2C isoenzymes are CYP2C8 and CYP2C9, present in roughly equal amounts and together comprising the vast majority of 2C proteins expressed (Furuya et al. 1991; Romkes et al. 1991). They are not polymorphic, however. The polymorphic isoenzyme appears to be CYP2C19 (Wrighton et al. 1993). There is a significant correlation between CYP2C19 level and (S)-mephenytoin 4'hydroxylase activity. CYP2C19 levels as determined by immunoblotting from 14 human liver samples exhibited a tenfold activity variability; with one liver found absent for activity, but which was also absent for detectable CYP2C19 isoenzyme. Conflicting data exist as to whether the closely related CYP2C18 isoenzyme is also a significant polymorphic 4'hydroxylase. CYP2C9 catalyses the 4'hydroxylation of (R)-mephenytoin, a nonpolymorphic reaction. Neither CYP2C9 nor CYP2C8 can metabolize (S)-mephenytoin, the polymorphic reaction (Relling et al. 1990; Veronese et al. 1993). Oral clearance measurements of the (S)-mephenytoin enantiomer in 13 subjects ranged from about 0.4 to 101/min, a range greater than 100-fold (Wedlund et al. 1985). Competitive inhibitors of the polymorphic (S)-mephenytoin 4'hydroxylase include diazepam, omeprazole, and proguanil.

The abundant liver isoenzyme CYP2C9 is the primary tolbutamide hydroxylase in man. CYP2C8 can participate to a small extent as well in tolbutamide metabolism. The CYP2C9 also appears to be chiefly responsible for phenytoin metabolism and is in addition the principal warfarin 7-hydroxylase in man. Warfarin hydroxylation as a tool is informative for evaluating the fine substrate specificities of the different 2C isoenzyme members (Kaminsky et al. 1993). CYP2C9 is regioselective for 7-hydroxywarfarin, with stereospecificity for (*S*)-warfarin; while CYP2C8 is also regioselective for 7-hydroxywarfarin but stereospecific for (*R*)-warfarin. CYP2C19 is regiospecific for 6- and 8-hydroxywarfarin and stereospecific for (*R*)-warfarin. CYP2C18 is regiospecific for 4-hydroxywarfarin but is without stereospecificity for warfarin as a substrate.

Variability of liver CYP2C8 isoenzyme has been measured at about tenfold, and interindividual variability of CYP2C9 at about fourfold, based on polymerase chain reaction determinations of specific mRNA level (Furuya et al. 1991). CYP2C18 is expressed at only low levels, reported to be between 2%–4% and 2%–10% of total CYP2C P450. Little is known about human 2C gene regulation. The CYP2C9 and CYP2C18 genes have multiple consensus sequences for glucocorticoid regulatory elements and one 15 base pair sequence with homology to the barbiturate-responsive element of P450$_{BM-3}$ in *Bacillus megaterium* (De Morais et al. 1993).

8.3.2.4 Human Isoenzyme P450 2D6

The CYP2D6 isoenzyme is famous for its genetic polymorphism, the debrisoquine/sparteine polymorphism of man. About 7% of the Caucasian population are poor metabolizers i.e., genetically deficient for CYP2D6 isoenzyme. The genetic defects and the clinical manifestations have been well documented (Meyer et al. 1990; Eichelbaum and Gross 1990). The list of drugs affected by this polymorphism now includes well over 20, ranging from β-adrenoreceptor antagonists, antiarrhythmics, tricyclic antidepressants, neuroleptics, and some morphine-related drugs. Apparently substrates metabolized by CYP2D6 tend not to be well metabolized by other major P450 isoenzymes.

Recently a novel but rare new type of CYP2D6 genetic polymorphism was uncovered after study of a small percentage of the Swedish population who were observed to be extraordinarily rapid metabolizers of debrisoquine (Johansson et al. 1993). Inherited amplification of the CYP2D6 gene was subsequently identified in these individuals, with some individuals containing up to 12 CYP2D6 gene copies. For these patients the clinical situation is reversed from that of the poor metabolizer phenotype, that is, there is the problem of reaching therapeutic drug concentration due to extensive metabolism. In individuals of normal CYP2D6 genotype there is nonetheless approximately a 16-fold variability of liver CYP2D6 isoenzyme level for reasons not well determined.

8.3.2.5 Human Isoenzyme P450 2E1

Chemical inducers of CYP2E1 include pyrazole, isoniazid, acetone, and also ethanol, which is probably the best known inducer and is responsible for its original name as the ethanol-inducible P450. Cyp2E1 is subject to regulation at more levels than other isoenzymes. These inducers to some extent can elevate CYP2E1 through increased mRNA stability and mRNA translation. The primary mechanism, however, is the ability of these inducers, which are for the most part also CYP2E1 substrates, to stabilize the enzyme (Kim et al. 1993). This stabilization appears to be a ligand-mediated mechanism, perhaps by protecting the enzyme from phosphorylation which may be a signal for protein degradation (Eliasson et al. 1990).

Substrates for CYP2E1 include acetaminophen and many low molecular weight solvents (Guengerich et al. 1991b). Chlorzoxazone has been proposed as a specific CYP2E1 substrate (Peter et al. 1990). A genetic polymorphism has been observed for the CYP2E1 gene but does not affect the coding region (Hayashi et al. 1991a,b; Persson et al. 1993). The polymorphic site described among Japanese is situated in the 5'-flanking region and is part of a putative HNF-1 binding site. It could conceivably contribute to the control of CYP2E1 gene expression, and the polymorphism may be associated with lung cancer risk. CYP2E1 is an abundant human liver isoenzyme; patient variability is moderate, reported at about fourfold (Forrester et al. 1992), but inducibility should be considered as likely especially as influenced by drinking. A selective microsomal CYP2E1 inhibitor is diethyldithiocarbamate (Guengerich et al. 1991a).

8.3.3 Family III Isoenzymes

8.3.3.1 Human Isoenzymes P450 3A4 and 3A5

In humans CYP3A4 and CYP3A5 are abundant liver P450s, representing together about 35% (range 17%–51%) of total hepatic P450 based on immunoblotting measurements (Wrighton et al. 1990). CYP3A5, when present, was found at levels less than CYP3A4, usually between 6% and 60% of total 3A family protein. In absolute terms CYP3A4 has been found at 6–260 pmol/mg microsomes (40-fold variation), and CYP3A5 between 2–60 pmol/mg microsomes (30-fold variation). CYP3A4 is responsible for the metabolism of many common drugs such as cyclosporine, diltiazem, lidocaine, and lovastatin (Kronbach et al. 1989; Pichard et al. 1990a,b; Bargetzi et al. 1989; Wang et al. 1991). CYP3A4 is inhibited in patients by flavonoids found in grapefruit (but not orange) juice (Guengerich and Kim 1990; Miniscalco et al. 1992). The erythromycin breath test has been proposed as an in vivo probe for measuring CYP3A4 activity (Watkins et al. 1992).

Probably the clinically most relevant inductions of P450 are from the induction of CYP3A4. This is due to the relative abundance of CYP3A4 in human liver, its ability to metabolize a broad range of drugs, and the range of chemicals and drugs capable of causing induction. CYP3A4 appears to be inducible in vivo following administration of antiseizure medications such as phenytoin and by the glucocorticoid analogue dexamethasone (Shaw et al. 1989). It appears that 3A protein levels in individuals treated with long-term anticonvulsant drugs are elevated typically about fivefold relative to untreated individuals. Furthermore, phenobarbital also appears to induce 3A4, indicating that this gene also contains phenobarbital control elements. CYP3A5, on the other hand, does not appear to be influenced by the administration of inducers of the 3A family. This situation appears analogous to rat PCN-2, which is constitutively expressed in male rats but is also not PCN inducible.

For CYP3A5 one study has reported that only 10%–20% of adult liver samples ($n = 40$) are positive for this isoenzyme, based on immunoblotting (Aoyama et al. 1989). A second report suggests a 29% incidence ($n = 60$; Wrighton et al. 1990). That means that the majority of persons express little of this isoenzyme. This isoenzyme alone may not play a critical or clinically significant role for drug metabolism, as, if symptomatic, such a variability would probably have been clinically uncovered in the past. Likewise CYP3A5 does not appear to be vital for endogenous, for examples, steroid function. It appears from in vitro data that most drug and steroid metabolic reactions carried out by CYP3A5 are also carried out by closely related isoenzyme CYP3A4 (Aoyama et al. 1989). This fact probably explains why the phenotypic relevance of CYP3A5 appears to have been concealed.

8.4 Molecular Biology: Recombinant Isoenzymes

It is now appreciated in both academia and industry that a highly attractive way to investigate the complex P450 family is to express the isoenzymes by gene technology. There is also little doubt of the importance that isoenzymes will play in pharmaceutical drug evaluation and development.

From the standpoint of different gene expression methods available, several have been used successfully for P450 heme-protein expression. The vaccinia virus has been used extensively for the expression of mammalian cytochrome P450 by the group of Gonzalez at NIH (Gonzalez et al. 1991a). A variety of rat and human isoenzymes have been expressed in Doehmer's laboratory using V79 Chinese hamster fibroblast cells genetically engineered for the stable expression of cytochrome P450 (Doehmer and Oesch 1991). The expression of a panel of human P450 isoenzymes has been achieved by the group of Crespi using Epstein-Barr virus based vectors transfected into human B lymphoblastoid cells (Crespi 1991). Using nonmammalian type hosts, recent progress has included *Escherichia coli* based P450 expression systems, pioneered by the group of Waterman (Barnes et al. 1991). *E. coli* based methods have been extended to cytochrome P450/P450 reductase fusion constructs, whereby a physically coupled and catalytically active mono-oxygenase is reconstructed from what otherwise are two independent proteins (Shet et al. 1993). Considerable success has been achieved by the group of Pompon in yeast, which are highly amenable for manipulation as to the coexpression of controlled amounts of cytochrome P450 reductase and cytochrome b_5 (Truan et al. 1993; Urban et al. 1993). Rounding off nonmammalian expression strategies also includes insect cell baculovirus P450 expression (Gonzalez et al. 1991b). There are additional mammalian cell expression strategies including those used for human CYP2E1 expression in HepG2 cells and in PC-12 cells (Dai et al. 1993; Mapoles et al. 1993). An innovative approach for P450 mutagenicity study has also recently included the development of immortalized normal human liver epithelial cells with

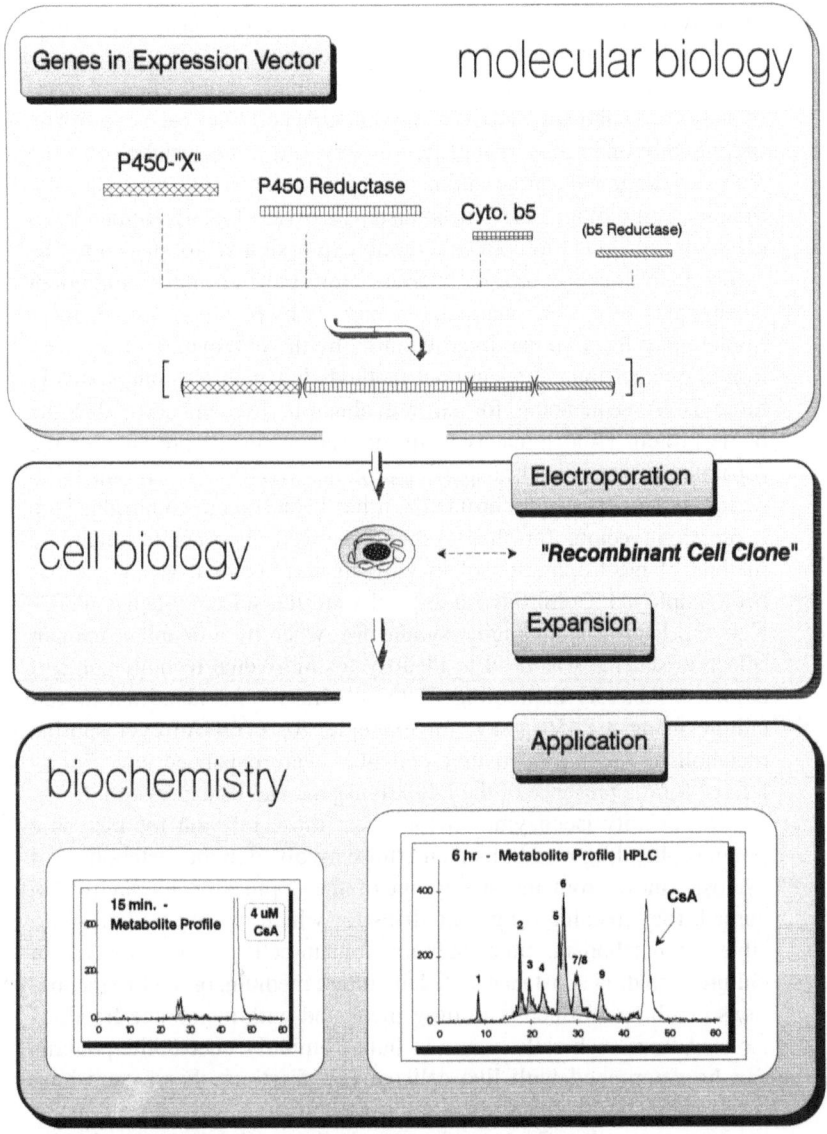

Fig. 4. Overview of molecular biology, cell biology, and biochemistry of development through application: recombinant isoenzymes of human cytochrome p450

partially retained hepatocyte characteristics and metabolism of chemical carcinogens (Pfeifer et al. 1993).

Our approach at Sandoz has aimed principally at developing a series of stable recombinant CHO-type mammalian cell lines each expressing individually one human isoenzyme involved in drug metabolism. The technical approach can be outlined briefly, which shows the molecular biology, cell biology, and biochemistry involved (Fig. 4). Human P450 cDNA is subcloned into an eukaryotic expression vector driven by the cytomegalovirus promotor. cDNAs for the 11 most important isoenzymes have been obtained, the majority by polymerase chain reaction cloning from human liver. Cloning by the polymerase chain reaction is quicker than conventional methods but requires comparatively more DNA sequencing for error evaluation. Remembering that the heme protein P450 is catalytically dependent on the presence of the separate electron chain transporting proteins NADPH-cytochrome P450 reductase and often cytochrome b_5, it has been tried to cointroduce the expression vectors for these protein as well. For cointroduction a method of electroporation of in vitro ligated vector concatamers has been employed. Cointroduced as well were the selection genes pSV2-Neo and bacterial asparagine synthetase, whereby a double-dominant selection scheme was used to identify for individual recombinant cell colonies. Actively expressing CHO cell lines were identified in cell culture using a P450 assay; for example, for CYP3A4 cyclosporine metabolism was used. Positive cell lines were expanded and characterized for the presence of the P450 transgene and its expression.

The priority isoenzymes for us were those relevant for assessing polymorphic drug metabolism and those involved in the metabolism of cyclosporines. From the standpoint of the application hierarchy discussed, for screening drug candidates for which isoenzyme is responsible for metabolism, either the recombinant cells can be employed in culture with drug substance added to culture medium, or in vitro incubations can be performed with drug incubated with membrane fractions isolated from such cells grown in bulk. Similarly, metabolite patterns can be determined with live cells or cell fractions. For establishing reaction kinetics the controlled in vitro incubations with isolated membranes are superior. Cyclosporine metabolism and drug interactions at the target of CYP3A4 have been well evaluated in vitro (Pichard et al. 1990a). For the small-scale generation of metabolites from CHO cells it has been possible to scale-up live cell incubations (less than 1l). Figure 4 (bottom) illustrates the metabolism of cyclosporine A by recombinant CYP3A4. Metabolite profiles of incubations with isolated cell membranes were taken after 15 min and 6 h. By 6 h both primary and secondary metabolites have formed.

Cells lines expressing the genetically polymorphic CYP2D6 have been used for assessing potential polymorphic metabolism of clozapine and tropisetron. The highly effective antipsychotic clozapine is subject to interindividual variability with regards to bioavailability, steady-state plasma concentrations, and clearance. The partial involvement of CYP2D6 in the overall metabolism of clozapine has been verified (Fischer et al. 1992). The recombinant isoenzyme identified CYP2D6-dependent metabolites which were centered around aromatic ring region hydroxylation, but CYP2D6 was not the only enzymatic pathway in man. The situation with the antiemetic serotonin type 3 antagonist tropisetron was more clearcut (Fischer et al. 1994). Relevant in vivo phenotyping had already suggested polymorphic CYP2D6 dependency for tropisetron metabolism. The catalytic activity of CYP2D6 was confirmed using the expressed isoenzyme. Metabolite patterns from expressed CYP2D6 showed three major hydroxylated metabolites. This metabolite profile matched that found in human plasma. In this case the major involvement of CYP2D6 in systemic metabolism was verified by a variety of methods. With regards to the interpretation of polymorphic drug metabolism it should not be considered incompatible with drug development but rather be seen for its advantages for further development. For improving clinical trial patient selection or dose-finding interpretation, individuals could be rationally screened, categorized, and properly interpreted in view of metabolizer phenotype. This is also the case for customized dosing scheme possibilities for drugs already on the market.

As the availability of human isoenzymes for use in pharmaceutical evaluation improves, other issues assume importance. It has been outlined how information can be related to practical value. Issues regarding the proper control and validation of recombinant enzymes for human drug metabolism investigation and a commentary on industrial application have recently been presented (Remmel and Burchell 1993; Tarbit et al. 1993). As we are learning more about the fundamental processes of human metabolite formation and extend our investigative abilities beyond mere metabolite identification, a database is growing in pharmaceutical research. With such a database and new investigative tools will come the increased responsibility for proper data presentation and acceptance both in-house and by regulatory authorities. This new progress is certainly visible to the regulatory authorities, who are either academically based or, at least, advised. The challenge from a pharmaceutical perspective is the proper and timely integration of this expanding database into the responsibility of maintaining excellence in pharmaceutical development.

8.5 Conclusions

1. In summary, the keyword in isoenzyme application for drug candidate evaluation would be a qualitative improvement in predictive capacity.
2. An "information hierarchy" in P450 isoenzyme application for pharmaceutical drug candidate evaluation and development is presented, along with a strategic outline coupling the information obtained with suggestions of its relevance for interpretation and value.
3. Multiple isoenzyme gene expression strategies are available for the pharmaceutical industry; each with its own particular strengths, but each apparently applicable for answering at least the basics of isoenzyme relevance for candidate drug evaluation and development.
4. The development and application of genetically engineered CHO-type mammalian cells has been our research focus, with an emphasis on those isoenzymes most clinically relevant for interindividual variabilities and those involved in cyclosporines metabolism.
5. The continued development of interpretive expertise is foreseen: for bridging isoenzyme data to whole cells, whole organs, and eventually in vivo, where ultimately the data can be inserted into phamacokinetic modeling.
6. The exploitation of a rapidly growing database of known isoenzyme substrates and drug interactions will continue to improve the significance for drug candidate evaluation.

7. With increased use of isoenzymes, future challenges include its integration in normal drug development schemes and its proper packaging and presentation for acceptance both in-house and by registration authorities.

References

Aoyama T, Yamano S, Waxman, DJ, Lapenson DP, Meyer UA, Fischer V, Tyndale R, Inabe T, Kalow W, Gelboin HV, Gonzalez FJ (1989) Cytochrome P-450 hPCN3, a novel cytochrome P-450 IIIA gene product that is differentially expressed in adult human liver. cDNA and deduced amino acid sequence and distinct substrate specificities of cDNA expressed hPCN1 and hPCN3 for the metabolism of steroid hormones and cyclosporine. J Biol Chem 264:10388–10395

Bargetzi MJ, Aoyama T, Gonzalez FJ, Meyer UA (1989) Lidocaine metabolism in human liver microsomes by cytochrome P450IIIA4. Clin Pharmacol Ther 46:521–527

Barnes HJ, Arlotto MP, Waterman MR (1991) Expression and enzymatic activity of recombinant cytochrome P450 17α-hydroxylase in escherichia coli. Proc Natl Acad Sci USA 88:5597–5601

Burbach KM, Poland A, Bradfield CA (1992) Cloning of the Ah-receptor cDNA reveals a distinctive ligand-activated transcription factor. Proc Natl Acad Sci USA 89:8185–8189

Butler MA, Iwasaki M, Guengerich FG, Kadlubar FF (1989) Human cytochrome P-450PA (P-4501A2), is primarily responsible for the hepatic 3-demethylation of caffeine and N-oxidation of carcinogenic arylamines. Proc Natl Acad Sci USA 86:7696–7700

Crespi CL (1991) Expression of cytochrome P450 cDNAs in human B-lymphoblastoid cells: application to toxicology and metabolite analysis. Methods Enzymol 206:123–130

Crofts F, Cosma GN, Currie D, Taioli P, Garte S (1993) A novel CYP1A1 gene polymorphism in African-Americans. Carcinogenesis 9:1729–1731

Dai Y, Rashbe-Step J, Cederbaum AI (1993) Stable expression of human cytochrome P4502E1 in HepG2 cells: characterization of catalytic activities and production of reactive oxygen intermediates. Biochemistry 32:6928–6937

Daujat M, Peryt B, Lesca P, Fourtanier G, Domergue J, Maurel P (1992) Omeprazole, an inducer of human CYP1A1 and 1A2, is not a ligand for the ah receptor. Biochem Biophys Res Commun 188:820–825

Dayer P, Desmeules J, Leeman T, Striberni R (1988) Bioactivation of the narcotic drug codeine in human liver is mediated by the polymorphic monooxygenase catalysing debrisoquine 4-hydroxylation (cytochrome P-450
db1/buff). Biochem Biophys Res Commun 152:411–416

De Morais SMF, Schweikl H, Blaisdell J, Goldstein JA (1993) Gene structure
and upstream regulatory regions of human CYP2C9 and CYP2C18.
Biochem Biophys Res Commun 194:194–201

Doehmer J, Oesch F (1991) V79 Chinese hamster cells genetically engineered
for stable expression of cytochromes P450. Methods Enzymol 206:117–123

Eichelbaum M, Gross AS (1990) The genetic polymorphism of debriso-
quine/sparteine metabolism – clinical aspects. Pharmacol Ther 46:377–394

Eliasson E, Johansson I, Ingelman-Sundberg M (1990) Substrate-, hormone-,
and cAMP-regulated cytochrome P450 degradation. Proc Natl Acad Sci
USA 87:325–329

Fischer V, Vogels B, Maurer G, Tynes RT (1992) The antipsychotic clozapine
is metabolized by the polymorphic human microsomal and recombinant
cytochrome P450 2D6. J Pharmacol Exp Ther 260:1355–1360

Fischer V, Vickers AEM, Heitz F, Mahadevan S, Baldeck J-P, Minery P,
Tynes R (1994) The polymorphic cytochrome P-4502D6 is involved in the
metabolism of both 5-hydroxytryptamine antagonists, tropisetron and on-
dansetron. Drug Metab Dispos 22 (in press)

Forrester LM, Henderson CJ, Glancey MJ, Back DJ, Park BK, Ball SJ, Kitter-
ingham NR, McLaren AW, Miles JS, Skett P, Wolf CR (1992) Relative expression of cytochrome P450 isoenzymes in human liver and association
with the metabolism of drugs and xenobiotics. Biochem J 281:359–368

Fujii-Kuriyama Y, Imataka H, Sogawa K, Yasumoto K-I, Kikuchi Y (1992)
Regulation of CYP1A1 expression. FASEB J 6:706–710

Furuya H, Meyer UA, Gelboin HV, Gonzalez FJ (1991) Polymerase chain re-
action-directed identification, cloning, and quantification of human
CYP2C18 mRNA. Mol Pharmacol 40:375–382

Gonzalez FJ (1990) Molecular genetics of the P-450 superfamily. Pharmacol
Ther 45:1–38

Gonzalez FJ (1992) Human cytochromes P450: problems and prospects. TIPS
13:346–352

Gonzalez FJ, Aoyama T, Gelboin (1991a) Expression of mammalian cytochrome P450 using vaccinia virus. Methods Enzymol 206:85–92

Gonzalez FJ, Kimura S, Tamura S, Gelboin (1991b) Expression of mammalian
cytochrome P450 using baculovirus. Methods Enzymol 206:93–99

Guengerich FP, Kim D-H (1990) In vitro inhibition of dihydropyridine oxidation and aflatoxin B1 activation in human liver microsomes by naringenin
and other flavonoids. Carcinogenesis 11:2275–2279

Guengerich FP (1992) Characterization of human cytochrome P450 Enzymes. FASEB J 6:745–748

Guengerich FP, Brian WR, Sari M-A, Ross JT (1991a) Expression of mammalian cytochrome P450 enzymes using yeast-based vectors. Methods Enzymol 206:130–148

Guengerich FP, Kim D-H, Iwasaki M (1991b) Role of human cytochrome P-450 IIE1 in the oxidation of many low molecular weight cancer suspects. Chem Res Toxicol 4:168–179

Hayashi S-I, Watanabe J, Kawajiri K (1991a) Genetic polymorphisms in the 5'-flanking region change transcriptional regulation of the human cytochrome P450IIE1 gene. J Biochem 10:559–565

Hayashi S-I, Watanabe J, Nakachi K, Kawajiri K (1991b) Genetic linkage of lung cancer-associated MSPI polymorphisms with amino acid replacement in the heme binding region of the human cytochrome P4501A1. Gene J Biochem 110:407–411

Johansson I, Lundqvist E, Bertilsson L, Dahl M-L, Sjöqvist F, Ingelman-Sundberg M (1993) Inherited amplification of an active gene in the cytochrome P450 CYP2D locus as a cause of ultrarapid metabolism of debrisoquine. Proc Natl Acad Sci USA 90:11825–11829

Kaminsky LS, de Marais SMF, Faletto MB, Dunbar DA, Goldstein JA (1993) Correlation of human cytochrome P4502C substrate specificities with primary structure: warfarin as a probe. Mol Pharmacol 43:234–239

Kim H, Putt D, Reddy S, Hollenberg PF, Novak RF (1993) Enhanced expression of rat hepatic CYP2B1/B2 and 2E1 by pyridine: differential induction kinetics and molecular basis of expression. J Pharmacol Exp Ther 267:927934

Kronbach T, Fischer V, Meyer UA (1989) Cyclosporine metabolism in human liver: identification of a cytochrome P-450III gene family as the major cyclosporine-metabolizing enzyme explains interactions of cyclosporine with other drugs. Clin Pharmacol Ther 43:630–635

Lambert GH, Schoeller DA, Humphry HE, Kotake AN, Lietz H, Campbell M, Kalow W, Spielberg SP, Budd M (1990) The caffeine breath test and caffeine urinary metabolite ratios in the michigan cohort exposed to polybromonated biphenyls: a preliminary study. Environ Health Perspect 89:175-181

Mapoles J, Berthou F, Alexander A, Simon F, Ménez J-F (1993) Mammalian PC-12 cell genetically engineered for human cytochrome P450 2E1 expression. Eur J Biochem 214:735–745

Mayumi M, Baba T, Yamazaki H, Ohmori S, Inui Y, Gonzalez FJ, Guengerich FP, Shimada T (1993) Characterization of cytochrome P-450 2B6 in human liver microsomes. Drug Metab Dispos 21:1048–1056

Meyer UA, Skoda RC, Zanger UM (1990) The genetic polymorphism of de-brisoquine/sparteine metabolism – molecular mechanisms. Pharmacol Ther 46:297–308

Mikus G, Somogyi AA, Bochner F, Eichelbaum M (1991) Codeine O-de-methylation: rat strain differences and the effects of inhibitors. Biochem Pharmacol 41:757–762

Miniscalco A, Lundahl J, Regårdh G, Edgar B, Eriksson UG (1992) Inhibition of dihydropyridine metabolism be flavonoids found in grape juice. J Pharmacol Exp Ther 261:1195–1200

Murray GI, Foster CO, Barnes TS, Weaver RJ, Snyder CP, Ewen SWB, Melvin WT, Burke MD (1992) Cytochrome P4501A1 expression in adult and fetal human liver. Carcinogenesis 13:165–169

Nelson DR, Kamataki T, Waxman DJ, Guengerich FP, Estabrook RW, Feyereisen R, Gonzalez FJ, Coon MJ, Gunsalus IC, Gotoh O, Okuda K, Nebert DW (1993) The P450 superfamily: update on new sequences, gene mapping, accession numbers, early trivial names of enzymes, and nomenclature. DNA Cell Biol 12:1–51

Omiecinski CJ, Redlich CA, Costa P (1990) Induction and developmental expression of cytochrome P4501A1 messenger RNA in rat and human tissues: detection by the polymerase chain reaction. Cancer Res 50:4315–4321

Pearce R, Greenway D, Parkinson A (1992) Species differences and interindividual variation in liver microsomal cytochrome P450 2A enzymes: effects on coumarin, dicumarol, and testosterone oxidation. Arch Biochem Biophys 298:211–225

Persson I, Johansson I, Bergling H, Dahl M-L, Seidegard J, Rylander R, Rannug A, Högberg J, Ingelman-Sundberg M (1993) Genetic polymorphism of cytochrome P4502E1 in a Swedish population. Relationship to incidence of lung cancer. FEBS Lett 319:207–211

Peter R, Böcker R, Beaune PH, Iwasaki M, Guengerich FP, Yang CS (1990) Hydroxylation of chlorzoxazone as a specific probe for human liver cytochrome P-450IIE1. Chem Res Toxicol 3:566–573

Pfeifer AMA, Cole KE, Smoot DT, Weston A, Groopman JD, Shields PG, Vignaud J-M, Juillerat M, Lipski MM, Trump BF, Lechner JF, Harris CC (1993) Simian virus 40 large tumor antigen-immortalized normal human liver epithelial cells express hepatocyte characteristics and metabolize chemical carcinogens. Proc Natl Acad Sci USA 90:5123–5127

Pichard L, Fabre I, Fabre G, Domergue J, Saint Aubert B, Mourad G, Maurel P (1990a) Cyclosporine A drug interactions: screening for inducers and inhibitors of cytochrome P-450 (cyclosporine A oxidase) in primary cultures of human hepatocytes and in liver microsomes. Drug Metab Dispos 18:595–606

Pichard L, Gillet G, Fabre I, Dalet-Beluche I, Bonfils C, Thenot J, Maurel P (1990b) Identification of the rabbit and human cytochromes P-450III as the major enzymes involved in the N-demethylation of diltiazem. Drug Metab Dispos 18:711–719

Relling MV, Aoyama T, Gonzalez FJ, Meyer UA (1990) Tolbutamide and mephenytoin hydroxylation by human cytochrome P450s in the CYP2C subfamily. J Pharmacol Exp Ther 252:442–447

Remmel RP, Burchell B (1993) Validation and use of cloned, expressed human drug-metabolizing enzymes in heterologous cells for analysis of drug metabolism and drug-drug interactions. Biochem Pharmacol 46:559–566

Romkes M, Faletto MB, Blaisdell JA, Raucy JL, Goldstein JA (1991) Cloning and expression of complementary DNAs for multiple members of the human cytochrome P450IIC subfamily. Biochemistry 30:3247–3255

Sesardic D, Boobis AR, Edwards RJ, Davies DS (1989) A form of cytochrome P450 in man, orthologous to form d in the rat, catalyses the O-deethylation of phenacetin and is inducible by cigarette smoking. Br J Clin Pharmacol 26:363–372

Sesardic D, Pasanen M, Pelkonen O, Boobis AR (1990) Differential expression and regulation of members fo the cytochrome P4501A gene family in human tissues. Carcinogenesis 11:1183–1188

Shaw PM, Barnes TS, Cameron D, Engeset J, Melvin WT, Omar G, Petrie JC, Rush WR, Snyder CP, Whiting WT, Wolf CR, Burke MD (1989) Purification and characterization of an anticonvulsant-induced human cytochrome P450 catalysing cyclosporine metabolism. Biochem J 263:653–663.

Shet MS, Fisher CW, Holmans PL, Estabrook RW (1993) Human cytochrome P450 3A4: enzymatic properties of a purified recombinant fusion protein containing NADPH-P450 reductase. Proc Natl Acad Sci USA 90:11748-11752

Shimida T, Yum C-H, Yamazaki H, Gautier J-C, Beaune PH, Guengerich FP (1992) Characterization of human lung microsomal cytochrome P450 1A1 and its role in the oxidation of chemical carcinogens. Mol Pharmacol 41:856–864

Tarbit MH, Bayliss MK, Herriot D, Hood SR, Hutson JL, Park GR, Serabjit-Singh CJ (1993) Applications of molecular biology and in vitro technology to drug metabolism studies: an industrial perspective. Biochem Soc Trans 21:1018–1023

Truan G, Cullen C, Reisdorf P, Urban P, Pompon D (1993) Enhanced in vivo monooxygenase activities of mammalian P450s in engineered yeast cells producing high levels of NADPH-P450 reductase and human liver cytochrome b5. Gene 125:49–55

Urban P, Truan G, Gautier J-C, Pompon D (1993) Xenobiotic metabolism in humanized yeast: engineered yeast cells producing human NADPH-cytochrome P-450 reductase, cytochrome b5, epoxide hydrolase and P-450s. Biochem Soc Transact 21:1028–1034

Veronese ME, Doecke CJ, Mackenzie PI, McManus ME, Miners JO, Rees DLP, Gasser R, Meyer UA, Birkett DJ (1993) Site-directed mutation studies if human liver cytochrome P-450 isoenzymes in the CYP2C family. Biochem J 289:533–538

Wang RW, Kari PH, Lu AYH, Thomas PE, Guengerich FP, Vyas KP (1991) Biotransformation of lovastatin. IV. Identification of cytochrome P450 3A proteins as the major enzymes responsible for the oxidative metabolism lovastatin in rat and human liver microsomes. Arch Biochem Biophys 290:355–361.

Watkins PB, Turgeon DJ, Saenger P, Lown KS, Kolars JC, Hamilton T, Fishman K, Guzelian PS, Voorhees JJ (1992) Comparison of urinary 6-β-cortisol and the erythromycin breath test as measures of hepatic P45-IIIA (CYP3A) activity. Clin Pharmacol Ther 266:265–273

Waxman DJ, Azaroff L (1992) Phenobarbital induction of cytochrome P-450 gene expression. Biochem J 281:577–592

Wedlund PJ, Aslanian WS, Jacqz E, McAllister CB (1985) Phenotypic differences in mephenytoin pharmacokinetics in normal subjects. J Pharmacol Exp Ther 234:662–669

Wrighton SA, Brian WR, Sarl M-A, Iwasaki M, Guengerich FP, Raucy JL, Molowa DT, Vandenbranden M (1990) Studies on the expression and metabolic capabilities of human liver cytochrome P450IIIA5 (HLp3). Mol Pharmacol 38:207–213

Wrighton SA, Stevens JC, Becker GW, Vandenbranden M (1993) Isolation and characterization of human liver cytochrome P450 2C19: correlation between 2C19 and S-mephenytoin 4'-hydroxylation. Arch Biochem Biophys 306:240–245

Yamano S, Nhamburo PT, Aoyama T, Meyer UA, Inabe T, Kalow W, Gelboin HV, McBride OW, Gonzalez FJ (1989) cDNA cloning and sequence and cDNA-directed expression of human P450 IIB1: identification of a normal and two variant cDNAs derived from the CTP2B locus on chromosome 19 and differential expression of the IIB mRNAs in human liver. Biochemistry 28:7340–7348

Yun C-H, Shimada T, Guengerich FP (1991) Purification and characterization of human liver microsomal cytochrome P-450 2A6. Mol Pharmacol 40:679–685

9 The Importance of Cytochrome P450 3A Enzymes in Drug Metabolism

F. P. Guengerich, E. M. J. Gillam, M. V. Martin, T. Baba,
B.-R. Kim, T. Shimada, K. D. Raney, and C.-H. Yun

9.1   Introduction 161
9.2   History of P450 3A 162
9.3   Regulation of P450 3A Family Enzyme Expression. 163
9.4   Expression Systems for P450 3A Enzymes 164
9.5   Roles of P450 3A4 in the Oxidation of Drugs 165
9.6   Roles of P450 3A4 in the Oxidation of Steroids.............. 168
9.7   Roles of P450 3A4 in the Oxidation of Carcinogens 170
9.8   In Vitro Methods of Identifying Roles for P450 3A4.......... 170
9.9   Noninvasive Assays for P450 3A Enzymes 171
9.10   Inhibition and Stimulation of P450 3A4.................... 172
9.11   Circulating Antibodies to P450 3A4....................... 173
9.12   Conclusions 175
References 176

9.1 Introduction

The biotransformation of drugs in the body has been appreciated for over a century (Keller 1842; Baumann and Preusse 1879; Jaffe 1879). Cytochrome P450 (P450) enzymes were reported more than 50 years ago (Mueller and Miller 1953; Brodie et al. 1958), and they are now recognized to be the major catalysts involved in the metabolism of drugs, steroids, carcinogens, pesticides, and pollutants (see entire January 1992 issue of *FASEB Journal*). There has been considerable

interest in being able to understand and predict transformation of xenobiotic chemicals in fields such as drug metabolism and chemical carcinogenesis. With appreciation of the increasing complexity of P450
enzymes in the 1970s it was unclear as to whether the biotransformation
of xenobiotics could ever be comprehended in terms of a limited number of P450 enzymes. In the 1980s there was extensive characterization
of many of the human P450s. Many drug oxidations can now be attributed to a few, major P450 enzymes. Several lines of evidence suggest that the enzyme P450 3A4 is, on the average, the major P450
present in human liver, and that small intestine plays a significant role in
the metabolism of many drugs.

9.2 History of P450 3A

The first account of a P450 3A enzyme was probably the report by Lu et
al. (1972) of selective induction of certain enzymatic activities by treatment of rats with the steroid pregnenolone 16α-carbonitrile. A protein
which was probably P450 3A1 was purified from livers of pregnenolone
16α-carbonitrile-treated rats by Elshourbagy and Guzelian (1980). P450
3A6 (then "P450 3c" and other terms) was purified from rabbits by
Ingelman-Sundberg et al. (1979), Miki et al. (1981), and Bonfils et al.
(1982).

   This laboratory purified a P450 protein responsible for oxidation of
the drug nifedipine from human liver (Guengerich et al. 1986a); antibodies raised to rat P450 3A or to some previously purified human liver
P450s (Wang et al. 1983) reacted with the protein (Guengerich et al.
1986a). The antibody was used to screen a cDNA library and isolate a
clone coding for the entire open reading frame (Beaune et al. 1986b).
This sequence and protein are now termed P450 3A4. An immunochemical screening approach was used by Watkins et al. (1985) to
isolate a human liver P450 cross-reactive with rat P450 3A enzymes. A
cDNA library was screened with an antibody to yield three fragments
that were overlapped to obtain a sequence now known as P450 3A3
(Molowa et al. 1986).

   Genomic blotting studies indicate that at least three genes exist in the
human P450 3A family (Beaune et al. 1986b). The number could be
higher, but it is our current opinion that all of the existing literature can

be rationalized in terms of three genes, P450s 3A4, 3A5, and 3A7. P450 3A3 (98% identical to 3A4) is a patched sequence (Molowa et al. 1986) and hepatic expression of 3A3 or another closely related clone, NF10, could not be detected in any of 12 human samples (Bork et al. 1989). P450 3A5 is polymorphically expressed; about one-fourth of individuals have the enzyme, and the level is usually approximately one-third that of P450 3A4 when it is expressed in liver (Aoyama et al. 1989; Wrighton and Vandenbranden 1989; Wrighton et al. 1989, 1990). Expression has also been reported in kidney (Watkins et al. 1992). The P450 3A7 protein (HLFa) and cDNA were first characterized in human fetal liver (Kitada and Kamataki 1979; Kitada et al. 1985; Komori et al. 1989). Komori et al. (1990) presented evidence that P450 3A7 mRNA was not found in adults although others have now reported expression in human endometrium and placenta (Schuetz et al. 1993).

9.3 Regulation of P450 3A Family Enzyme Expression

Clearly P450 3A4 is the most abundant of the P450 3A proteins in adult human liver (Bork et al. 1989; Komori et al. 1990). The level varies at least 40-fold (Guengerich 1988; Bork et al. 1989) and can rise to as high as 60% of the total P450 in a human liver (Guengerich 1990). Immunochemical evidence has been presented that the average level of P450 3A4 is higher than that of any other P450 (approx. 30% of total P450; Shimada et al. 1994). As pointed out above, levels of expression of P450 3A5 and 3A7 are usually considerably less. Evidence has also been presented that P450 3A4 is the major P450 3A enzyme (and P450) expressed in human small intestine.

The level of P450 3A enzymes in human liver has been shown to be elevated by treatment with barbiturates, rifampicin, and dexamethasone (Watkins et al. 1985). These changes have also been shown in hepatocyte culture (Morel et al. 1990). The induction of P450 3A4 can dramatically enhance the elimination of drugs and lead to lack of efficacy, as demonstrated in the case of the oral contraceptive 2-ethynylestradiol (Bolt et al. 1975, 1977; Guengerich 1988). Other human P450 enzymes can be induced by barbiturates (Morel et al. 1990; Zilly et al. 1975); however, the situation in humans probably contrasts considerably with that in experimental animals. In rats and rabbits, barbiturates produce strong induction of P450 2B enzymes. However, in all of 60 human liver samples examined the level of expression of P450 2B6 was considerably less than that of P450 3A4 (Mimura et al. 1993). The level of P450 3A4 tends to be reduced in cirrhosis (Guengerich and Turvy 1991; Kleinbloesem et al. 1986; Ene and Roberts 1987).

The mechanisms of regulation of P450 3A enzymes are still unclear. Evidence has been presented in experimental animal systems that dexamethasone induction does not involve the "classic" glucocorticoid receptor (Schuetz and Guzelian 1984). Some motifs for regulatory proteins have been observed in the 5' upstream regions (Itoh et al. 1992).

9.4 Expression Systems for P450 3A Enzymes

Several systems have now been used for the heterologous expression human P450 3A enzymes. Aoyama et al. (1989) expressed both P450 3A4 and 3A5 in HepG2 cells using vaccinia virus and measured catalytic activities towards nifedipine, benzo(a)pyrene, 7-ethoxycoumarin, testosterone, and cyclosporin A. P450 3A4 has also been stably expressed in human AHH-1 cells (Crespi et al. 1991). Kitamura et al. (1992) have expressed P450 3A7 in MCF7 human breast cancer cells.

Our laboratory expressed P450 3A4 in the yeast *Saccharomyces cerevisiae* and partially purified the enzyme for use in catalytic studies (Brian et al. 1990). Renaud et al. (1990) also expressed P450 3A4 in yeast. One of the problems that plagues work with the P450 3A enzymes is the difficulty in achieving expected catalytic activities in reconstituted enzyme systems (Elshourbagy and Guzelian 1980; Schwab et al. 1988; Guengerich et al. 1986a). Activities can be considerably higher in the presence of certain lipids and cytochrome b_5 (Imaoka et al. 1992; Halvorson et al. 1990). Pompon and his associates have integrated human NADPH-P450 reductase and cytochrome b_5 into the *S. cerevisiae* genome to produce a system in which the yeast microsomes have considerably better catalytic activities (Peyronneau et al. 1992).

Two approaches have been used for the expression of P450 3A4 in the bacterium *Escherichia coli*. We made a number of changes in the 5' terminus of the cDNA (in the pCW vector) and analyzed expression (Fig. 1). One of these constructs (NF14) yielded considerably more P450 3A4 than any of the others (Gillam et al. 1993). The level of

```
     1      5         10        15        20        25        30        35
NF1:  M A L I P D L A M E T W L L L A V S L V L L  Y L Y G T H S H G L F K K
NF10: M A ─────────────────────────────────────── Y G T H S H G L F K K
NF12: M A ─────────────── L L L A V ─────────────────────────── F K K
NF13: M A L I P D L A M E T W L L L A V S L V L L Y  L Y G T H S H G L F K K
NF14: M A ─────────────── L L L A V F L V L L  Y L Y G T H S H G L F K K
```

Fig. 1. Modifications of N-terminal sequence of P450 3A4 used in expression studies with the vector pCW in *Escherichia coli*. "NF1" denotes the native sequence (the terminal Met is cleaved in the *E. coli*). Only the NF14 sequence is expressed at a high level (Gillam et al. 1993)

expression is routinely about 100 nmol P450 per liter of culture, as recovered in the membrane fraction (10^5 g). P450 3A4 has been purified to near homogeneity in a simple two-step procedure and its spectral and catalytic properties are those expected (Gillam et al. 1993). As in the case of the purified hepatic enzyme, reconstitution of catalytic activity required cytochrome b_5, an unusual mixture of lipids, and the tripeptide glutathione (Gillam et al. 1993). The purified protein has been used to produce polyclonal antibodies in rabbits.

Another approach to bacterial expression has been employed by Fisher et al. (1992) and Shet et al. (1993), who fused the catalytic domain of rat NADPH-P450 reductase to the 3' end of P450 3A4 to generate a fusion protein that they were able to purify. The fusion protein had catalytic activity in the absence of added NADPH-P450 reductase. Surprisingly, lipids were still required for some catalytic activities (e.g., testosterone 6-β-hydroxylation) but not others (erythro-mycin N-demethylation; Shet et al. 1993).

9.5 Roles of P450 3A4 in the Oxidation of Drugs

P450 3A4 was isolated from human liver microsomes on the basis of its ability to catalyze the oxidation of the calcium channel blocker nife-dipine (Guengerich et al. 1986a). Since then the enzyme has been demonstrated to play a major role in the oxidation of a large number of other 1,4-dihydropyridines (Table 1) and other drugs (Table 2). It should be emphasized that many new drug substrates for P450 3A4 are being revealed in the process of systematic studies in the pharmaceutical industry and that the list in Table 2 is expected to grow (steroid drugs

Table 1. Dihydropyridine substrates for P450 3A4 (ring dehydrogenation in all cases) (Böcker and Guengerich; 1986, Guengerich et al. 1991, Brian et al. 1990)

Nifedipine
Nitrendipine
Niludipine
Nisoldipine
Nimodipine
Nicardipine
Felodipine
Bayer R4407 [(+)K8644]
Bayer R5417 [(-)K8644]
1,4-Dihydro-2,6-dimethyl-4-(3-nitro-4-chlorophenyl)-3,5-pyridinedicarboxylic acid dimethyl ester
1,4-Dihydro-2,6-dimethyl-4-phenyl-3,5-pyridinedicarboxylic acid dimethyl ester
1,4-Dihydro-2,6-dimethyl-4-phenyl-3,5-pyridinedicarboxylic acid dimethyl ester
1,4-Dihydro-2,6-dimethyl-4-(1-naphthalenyl)-3,5-pyridinedicarboxylic acid dimethyl ester
1,4-Dihydro-2,6-dimethyl-4-(3-chlorophenyl)-3,5-pyridinedicarboxylic acid dimethyl ester
1,4-Dihydro-2,6-dimethyl-4-(2,3,5,6-tetrafluorophenyl)-3,5-pyridinedicarboxylic acid dimethyl ester
1,4-Dihydro-2,6-dimethyl-4-[(4-trifluoromethyl)phenyl]-3,5-pyridinedicarboxylic acid dimethyl ester
1,4-Dihydro-2,6-dimethyl-4-(2-methyoxyphenyl)-3,5-pyridinedicarboxylic acid diethyl ester
1,4-Dihydro-2,6-dimethyl-4-(3-cyanophenyl)-3,5-pyridinedicarboxylic acid dimethyl ester
1,4-Dihydro-2,6-dimethyl-4-ethyl-3,5-pyridinedicarboxylic acid diethyl ester
1,4-Dihydro-2,6-dimethyl-4-benzyl-3,5-pyridinedicarboxylic acid dimethyl ester
1,4-Dihydro-2,6-dimethyl-4-(4-methylphenyl)-3,5-pyridinedicarboxylic acid dimethyl ester
1,4-Dihydro-2,6-dimethyl-4-(2-chloro-6-fluorophenyl)-3,5-pyridinedicarboxylic acid dimethyl ester

Table 2. Drug substrates for P450 3A4 (exclusive of dihydropyridines and steroids) (positions of oxidation are indicated in parentheses when available)

Substrate	Reference
Warfarin (R–10, S-dehydro)	Brian et al. 1990
Quinidine (3, N)	Guengerich et al. 1986b; Brian et al. 1990
Terfenadine (CH_3, N-dealkylation)	Yun et al. 1993
Cyclosporin A (AM9, AM1, AM4N; nomenclature formerly M1, M17, M21)	Kronbach et al. 1988; Combalbert et al. 1989; Aoyama et al. 1989
Midazolam (1, 4)	Kronbach et al. 1989
Triazolam	Kronbach et al. 1989
Lovastatin (6'β, 6'-*exo*-methylene, 3")	Wang et al. 1991
Alfentanil (noralfentanil)	Yun et al. 1992c
Lidocaine (N-deethylation	Bargetzi et al. 1989; Imaoka et al. 1990
Sulfentanil (N-dealkylation)	Tateishi et al. 1994
FK506	Sattler et al. 1992; Vincent et al. 1992
Rapamycin (41, others)	Sattler et al. 1992
Erythromycin (N-demethylation)	Watkins et al. 1985; Watkins et al. 1989; Brian et al. 1990
Benzphetamine (N-demethylation)	Guengerich et al. 1986a
Troleandomycin (N)	Renaud et al. 1990
Dapsone (N)	Fleming et al. 1992
Imipramine (N-demethylation)	Lemoine et al. 1993
Taxol (2-phenyl)	Cresteil et al. 1994; Harris et al. 1994
Codeine (N-demethylation)	Caraco et al., in preparation
Acetaminophen (quinone formation)	Patten et al. 1993
Aldrin (*exo* epoxidation)	Guengerich et al. 1986a
Omeprazole	Curi-Pedrosa et al. 1993; Birkett et al. 1993

are treated below.) In most of the cases P450 3A4 appears to be the dominant catalyst in many liver samples. In individuals with low levels of P450 3A4, other enzymes may be expected to contribute more.

Analysis of the list of P450 3A4 substrates shows essentially nothing in the way of a pharmacophore model for the enzyme, in contrast to P450 2D6 (Islam et al. 1991; Koymans et al. 1992; Strobl et al. 1993). Further, many of the substrates are very large (e.g., cyclosporin A) but in the smaller substrates there is considerable regio- and stereoselectivity (e.g., testosterone). Single electron transfer reactions may be involved with some of the amines and other low-potential substrates, but this mechanism (which is effective over long distances than hydrogen atom abstraction; Grace et al. 1994; Macdonald et al. 1989) is not general for all P450 3A substrates.

Other human P450 3A family enzymes have not been examined in such extensive detail. Recombinant and purified hepatic P450 3A5 proteins have been found to catalyze oxidations of nifedipine (Aoyama et al. 1989; Wrighton et al. 1990), and cyclosporin A (Aoyama et al. 1989) but not erythromycin or quinidine (Wrighton et al. 1990). P450 3A7 has been shown to oxidize benzphetamine, aminopyrine, and ethylmorphine (Kitada et al. 1985).

9.6 Roles of P450 3A4 in the Oxidation of Steroids

Following is a list of steroids for which P450 3A4 plays a major role in oxidation (Guengerich et al. 1986a; Waxman et al. 1988, 1991; Guengerich 1988, 1990; Ged et al. 1989; Brian et al. 1989):

– Testosterone (6β, trace 15β, 2β)
– Δ^4-Androstendione (6β)
– Cortisol (6β)
– Progesterone (6β, some 16α)
– 17β-Estradiol (2,4)
– 17α-Ethynylestradiol (2)
– Dehydroepiandrosterone 3-sulfate (16a)

The major site of hydroxylation is often 6β or 16α in the androgens and mineralocorticoids or the 2 position in estrogens. P450s 3A4 and 3A7 have been shown to catalyze dehydroepiandrosterone 3-sulfate

Table 3. Carcinogenic substrates for P450 3A4 (positions of oxidation indicated in parentheses when known from Guengerich and Shimada 1991)

Substrate	Reference
Aflatoxin B_1 (8,9; 3α)	Shimada and Guengerich 1989; Raney et al. 1992; Shimada et al. 1989a
Aflatoxin G1 (9,10)	Shimada and Guengerich 1989; Raney et al. 1992; Shimada et al. 1989a
Sterigmatocystin (2,3)	Shimada and Guengerich 1989; Raney et al. 1992; Shimada et al. 1989a
7,8-Dihydroxy-7,8-dihydro-benzo(a)pyrene (9,10)	Shimada et al. 1989a,b
6-Aminochrysene	Yamazaki et al. 1993
6-Nitrochrysene (NO_2 reduction)	Chae et al. 1993
1-Nitropyrene	Howard et al. 1990
Tris(2,3-dibromopropyl)phosphate	Shimada et al. 1989a
Senecionine (N, 1)	Miranda et al. 1991
Benzo(a)pyrene (3)	Yun et al. 1992b
4,4'-methylene-bis(2-chloraniline) (N)	Yun et al. 1992a
9,10-Dihydroxy-9,10-dhydro-benzo(b)fluoranthene	Shimada et al. 1989b
3,4-Dihydroxy-3,4-dihydro-7,12-diemthylbenz(a)anthracene	Shimada et al. 1989b
1,6-Dinitropyrene (NO_2 reduction)	Shimada and Guengerich 1990

16α-hydroxylation (Kitada et al. 1987; Brian et al. 1990). P450 3A5 has low activity in the oxidation of testosterone, Δ^4-androstenedione, progesterone, and cortisol and appreciable activity towards 17β-estradiol (Wrighton et al. 1990; Aoyama et al. 1989). However, no activity towards 17α-ethynylestradiol was detected (Wrighton et al. 1990).

9.7 Roles of P450 3A4 in the Oxidation of Carcinogens

A list of carcinogenic substrates of P450 3A4 is presented in Table 3. In some cases the role of P450 3A4 may be one of detoxication instead of activation. For instance, aflatoxin B_1 can be activated by 8,9-epoxidation (to the *exo* isomer) or inactivated by 3α-hydroxylation (Shimada and Guengerich 1989; Raney et al. 1992). Further, even 8,9-epoxidation of dietary aflatoxin in the small intestine might serve to protect an individual from hepatocarcinogenesis by eliminating the carcinogen.

P450 3A5 appeared to be considerably less effective than 3A4 in the activation of a number of carcinogens (Wrighton et al. 1990). However, P450 3A7 has been shown to be competent in many of these activations (Kitada et al. 1989, 1990). Since P450 1A2 is absent in human fetal liver, P450 3A7 plays a substantial role in the activation of many carcinogenic arylamines and heterocyclic amines in that situation (Kitada et al. 1990) (these compounds are not included in the list of Table 3).

9.8 In Vitro Methods for Identifying Roles for P450 3A4

Demonstrating a role for P450 3A4 in a particular reaction is not particularly difficult in vitro (Guengerich and Shimada 1991). If human liver samples are available, the correlation of rates of a reaction under consideration may be correlated to a marker in different liver samples (e.g., nifedipine oxidation or immunochemically determined levels of P450 3A4; Beaune et al. 1986a). In principle, the R_2 coefficient reflects the percentage of the variance that can be accounted for by the relationship (i.e., attributed to the enzyme). Another possibility is to examine the extent of inhibition of a reaction in a crude sample (i.e., microsomes) that contains all P450s or other enzymes capable of catalyzing the reaction (Thomas et al. 1977). P450 3A4 can be inhibited with a chemical (troleandomycin or gestodene, see above) or an antibody. In principle, the extrapolated maximum extent of inhibition reflects the percentage of the reaction attributed to the enzyme, if the inhibition is specific. The principal chemical inhibitors used for this purpose are troleandomycin, gestodene, and ketoconazole (see above). Another approach is to use purified or, more commonly now, expressed P450 3A4; while an activity may be demonstrated with such an isolated P450, it is important to (a) compare the activity to that of other P450 enzymes and (b) to consider the relative levels of expression of the different P450s in the liver or other tissue under consideration (as well as in an expression system itself).

9.9 Noninvasive Assays for P450 3A Enzymes

It is certainly of interest to use drugs to estimate levels of P450 3A enzymes (especially hepatic P450 3A4). A number of assays have been utilized but all have some attendant problems. The list includes nifedipine oxidation, erythromycin N-demethylation, lidocaine N-deethylation, midazolam 1- and 4-hydroxylation, dapsone N-hydroxylation, and cortisol 6β-hydroxylation.

A report of variation of in vivo nifedipine oxidation (Kleinbloesem et al. 1984) originally led us to consider the isolation of the enzyme. The level of activity shows variation of an order of magnitude in clearance or area-under-the-curve (Kleinbloesem et al. 1984; Schellens et al. 1988). The major drawback is the complexity of the assay, as further oxidation of the primary product occurs and the pharmacokinetics cannot be readily evaluated without repetitive plasma sampling.

Erythromycin N-demethylation seems to discriminate between P450 3A4 and 3A5 (Watkins et al. 1989; Wrighton et al. 1990). Radioisotopic material is used and must be given intravenously, thus requiring special procedures and approval (Watkins et al. 1989).

Lidocaine N-deethylation has been utilized primarily as an index of liver function prior to transplant (MEGX test; Oellerich et al. 1990). It has a potential drawback in that high levels of lidocaine need to be administered and can have pharmacological effects.

Cortisol 6β-hydroxylation was first used as a general means of assessing hepatic enzyme induction (Ohnhaus and Park 1979); later Ged et al. (1989) showed the role of P450 3A4 in the hydroxylation. This assay has the advantage that no drugs need to be administered. Renal production of the product has been cited as a confounder (Watkins et al. 1992). A radioimmunoassay kit is commercially available, but for accurate results the ratio of 6β-hydroxycortisol:cortisol should be used. 6α-Hydroxycortisol is present at substantial levels in many individuals and care should be taken that this does not interfere (J.D. Groopman, personal communication).

Midazolam and dapsone hydroxylation have been utilized in analysis of P450 3A4 (Kronbach et al. 1989; Fleming et al. 1992). To date they have not shown as much interindividual variation as some of the other assays.

Some of the disadvantages of the current assays have been pointed out. One troublesome general problem has been that the parameters measured with the different assays have shown poor agreement to date. Rates of all of these reactions tend to increase when barbiturates and other inducers are administered and impaired in cirrhosis (when examined to date; Ohnhaus and Park 1979; Regårdh et al. 1989; Park 1981; Kleinbloesem et al. 1986; Ene and Roberts 1987). However, the correlation between the different parameters in sets of healthy volunteers has not been good. Although the advantages of the erythromycin assay over 6β-hydroxycortisol excretion have been touted (Watkins et al. 1992), agreement with other parameters has been poor (A.J.J. Wood, G.R. Wilkinson, personal communication). Possible sources of disagreement include variability in levels of expression of P450 3A4 vs. 3A5, variable roles of intestinal P450 3A4 in the metabolism of some orally administered drugs, the influence of blood flow due to first-pass clearance, and low interindividual variability in some cases. While most of these assays are useful, there is not yet common agreement on a single assay, and validation is still an issue.

9.10 Inhibition and Stimulation of P450 3A Enzymes

Most of the work has been done with P450 3A4 and neither 3A5 nor 3A7 have been considered as extensively. Some of the irreversible, mechanism-based inactivators of P450 3A4 are troleandomycin and gestodene (Pessayre et al. 1983; Guengerich 1990). The in vivo effects of troleandomycin and related macrolide antibiotics such as erythromycin have been documented (Tinel et al. 1989); the in vivo effect of low doses of gestodene is controversial (Jung-Hoffmann and Kuhl 1990). Ketoconazole is an apparently competitive inhibitor and effects are seen in vivo. However, the ketoconazole inhibition is not totally specific (Vanden Bossche 1992) and with erythromycin sometimes inhibition is not seen in vitro and in vivo (Wang et al. 1991). Grapefruit juice can have a dramatic inhibitory effect on the oxidation of nifedipine and felodipine (orange juice does not; Bailey et al. 1991). The major flavonoid, naringenin, is somewhat inhibitory (Guengerich and Kim 1990) but may not account for the effect in vivo (Bailey et al. 1993). P450 3A4 is also sensitive to cimetidine inhibition (Knodell et al. 1991).

One of the more intriguing aspects of P450 3A4 is the direct stimulation seen with some chemicals. For instance, 7,8-benzoflavone (α-naphthoflavone) can enhance reactions such as aflatoxin B_1 8,9-epoxidation or benzo(a)pyrene 3-hydroxylation five- to tenfold (Shimada and Guengerich 1989; Shimada et al. 1989a). This observation explains the earlier observations of Conney and his associates (Buening et al. 1978, 1981) and is in stark contrast to the potent inhibition of all activities of P450 1A2 by the same compounds (Butler et al. 1989; McManus et al. 1990). However, some P450 3A4 reactions are refractory to 7,8-benzoflavone and some are even inhibited (e.g., alfentanil oxidation; Yun et al. 1992c). Even with the single substrate aflatoxin B_1 one reaction, 8,9-epoxidation, is stimulated while another, 3α-hydroxylation, is inhibited (Raney et al. 1992). This behavior suggests a possible allosteric interaction, as is the case with the rabbit ortholog P450 3A6 (Schwab et al. 1988). Also, aflatoxin B_2 inhibits aflatoxin B_1 8,9-hydroxylation but not 3α-hydroxylation (Raney et al. 1992). The substrate nifedipine inhibits both of the reactions of aflatoxin B_1 but only 3α-hydroxylation is inhibited by the oxidation product of nifedipine (Guengerich et al. 1993).

A possible model for the enzyme has an allosteric site distinct from the substrate binding site. Some substrates may bind to both. The validity of this hypothesis is under investigation.

9.11 Circulating Antibodies to P450 3A4

In at least five cases, circulating antibodies to human P450 enzymes are associated with disease (P450s 1A2, 2C9, 2D6, 17A, 21A; Bourdi et al. 1990; Beaune et al. 1987; Zanger et al. 1988; Krohn et al. 1992; Bednarek et al. 1992). In two cases the disease (hepatitis) is drug-related and the covalent attachment of a drug to the P450 is thought to be causally linked to disease (P450s 1A2, 2C9, Bourdi et al. 1990; Beaune et al. 1987). We considered this possibility with regard to some of the toxicity seen with cyclosporin A, a P450 3A4 substrate (Combalbert et al. 1989; Kronbach et al. 1988).

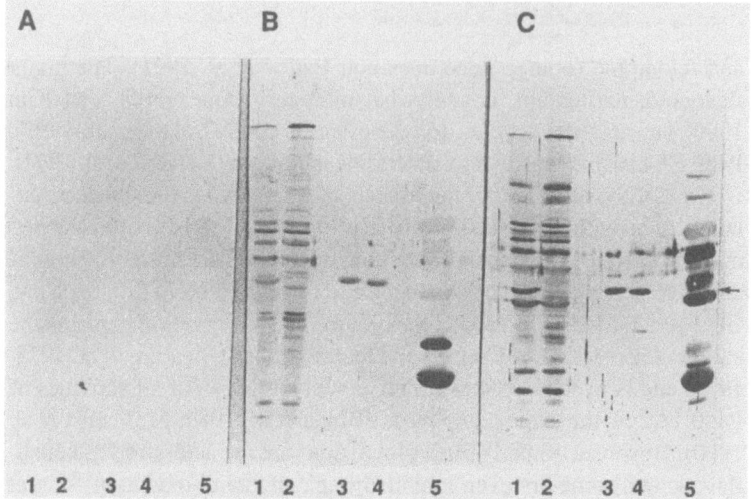

Fig. 2A–C. Immunochemical reaction of human proteins with antisera of patients receiving cyclosporin A therapy. Proteins were resolved by sodium dodecyl sulfate–polyacrylamide gel electrophoresis and electrophoretically transferred to nitrocellulose membranes (Guengerich et al. 1982). Sheets were blocked by treatment with phosphate-buffered saline (15 mM sodium phosphate buffer, pH 7.4, containing 0.15 M NaCl) and 0.55% Tween 20 (v/v). Sera were diluted 1/100 in the buffer containing 0.05% Tween 20 (v/v). Secondary antisera were diluted in phosphate-buffered saline buffer containing 0.05% Tween 20 (v/v), and in all cases each sheet was washed six times with this buffer between changes of reagents. The developing antibody was a 1/(5 × 10³) dilution of Pierce goat anti-human immunoglobulin G (H+L)/horseradish peroxidase conjugate (Rockford, IL). **B, C** Blots were treated, respectively, with 1/100 dilutions of the sera from patients I-B and I-C (4°C, overnight) prior to development with the secondary antibody system. The *Arrow (right)*, the position of P450 3A4. The three blots were developed and photographed at the same time. In each panel the individual lanes of the electrophoresis gel contained: *lane 1*, human liver microsomal sample HL 110 (23 μg protein); *lane 2*, human liver microsomal sample HL 126 (23 μg protein); *lane 3*, P450 3A4 purified from human liver (5 pmol = 0.3 μg protein); *lane 4*, recombinant P450 3A4 purified from *E. coli* (5 pmol = 0.3 μg protein); *lane 5*, standard human proteins (5 μg protein for each) serum albumin (65 kDa), epoxide hydrolase (50 kDa), lactic dehydrogenase (35 kDa), carbonic anhydrase (29 kDa), and glutathione S-transferase π (26 kDa). In **A** no primary antisera was used, and the blot was developed only with the secondary antibody system. In all cases color was developed with H_2O_2 and 4-chloro-1-naphthol

Sera from some heart transplant patients receiving cyclosporin A contained immunoglobulin G antibodies that reacted with hepatic or recombinant P450 3A4 (Fig. 2), as well as several other human proteins, under the conditions used. However, sera from some healthy volunteers who had never used cyclosporin also showed reactive antibodies, and the mean level of reactivity was not significantly different in the two groups ($n = 38$ for each group, $p > 0.10$).

We also demonstrated the in vitro conversion of (recombinant) P450 3A4 to a higher M_r form by apparent adduction of cyclosporin A in incubations. The apparent P450 3A4-cyclosporin A adducts were recognized by the human sera (that recognized P450 3A4) or anti-cyclosporin A but, surprisingly, not by rabbit anti-P450 3A4 antisera raised against the enzyme.

The clinical significance of these observations is not clear. However, in this case it does not appear that the presence of anti-P450 antibodies is associated with disease, in contrast to other P450s considered to date. It is conceivable that P450 3A4 often forms conjugates with common drugs that healthy volunteers might encounter (e.g., acetaminophen, erythromycin) to generate antigenic derivatives.

9.12 Conclusions

Investigations in several laboratories indicate that P450 3A4 is generally the most abundant P450 enzyme in human liver and small intestine. It is known to be inducible by barbiturates, dexamethasone, and rifampicin. The enzyme appears to play a significant role, at least partly due to its relative abundance, in the oxidations of many drugs. Many steroids and carcinogens are also substrates. Current knowledge suggests a large binding site, or perhaps an allosterically controlled one, that can show unusual behavior with substrates, inhibitors, and stimulators. It is possible to understand some untoward drug-drug interactions in terms of this enzyme. Methods are becoming available for in vitro and in vivo studies with new drug candidates and with individual humans. Circulating antibodies to P450 3A4 appear not to be related to disease states. P450 3A4 will continue to be one of the most significant enzymes under consideration in the pharmaceutical industry.

References

Aoyama T, Yamano S, Waxman DJ, Lapenson DP, Meyer UA, Fischer V, Tyndale R, Inaba T, Kalow W, Gelboin HV, Gonzalez FJ (1989) Cytochrome P-450 hPCN3, a novel cytochrome P-450 IIIA gene product that is differentially expressed in adult human liver. J Biol Chem 264:10388–10395

Bailey DG, Spence JD, Munoz C, Arnold JMO (1991) Interaction of citrus juices with felodipine and nifedipine. Lancet 337:268–269

Bailey DG, Arnold JMO, Munoz C, Spence JD (1993) Grapefruit juice-felodipine interaction: mechanism, predictability, and effect of naringin. Clin Pharmacol Ther 53:637–642

Bargetzi MJ, Aoyama T, Gonzalez FJ, Meyer UA (1989) Lidocaine metabolism in human liver microsomes by cytochrome P450IIIA4. Clin Pharmacol Ther 46:521–527

Baumann E, Preusse C (1879) Über Bromphenylmercaptursäure. Ber Dtsch Chem Ges 12:806

Beaune P, Kremers PG, Kaminsky LS, de Graeve J, Guengerich FP (1986a) Comparison of monooxygenase activities and cytochrome P-450 isozyme concentrations in human liver microsomes. Drug Metab Dispos 14:437–442

Beaune PH, Umbenhauer DR, Bork RW, Lloyd RS, Guengerich FP (1986b) Isolation and sequence determination of a cDNA clone related to human cytochrome P-450 nifedipine oxidase. Proc Natl Acad Sci USA 83:8064-8068

Beaune P, Dansette PM, Mansuy D, Kiffel L, Finck M, Amar C, Leroux JP, Homberg JC (1987) Human anti-endoplasmic reticulum autoantibodies appearing in a drug-induced hepatitis are directed against a human liver cytochrome P-450 that hydroxylates the drug. Proc Natl Acad Sci USA 84:551–555

Bednarek J, Furmaniak J, Wedlock N, Kiso Y, Baumann-Antczak A, Fowler S, Krishnan H, Craft JA, Smith BR (1992) Steroid 21-hydroxylase is a major autoantigen involved in adult onset autoimmune Addison's disease. FEBS Lett 309:51–55

Birkett DJ, Veronese ME, Miners JO, Andersson T (1993) In vitro characterisation of the human metabolism of omeprazole. Clin Exp Pharm Physiol [Suppl] 1:7

Böcker RH, Guengerich FP (1986) Oxidation of 4-aryl- and 4-alkyl-substituted 2,6-dimethyl-3,5-bis(alkoxycarbonyl)-1,4-dihydropyridines by human liver microsomes and immunochemical evidence for the involvement of a form of cytochrome P-450. J Med Chem 29:1596–1603

Bolt HM, Kappus H, Bolt M (1975) Effect of rifampicin treatment on the metabolism of oestradiol and 17α-ethinyloestradiol by human liver microsoms. Eur J Clin Pharmacol 8:301–307

Bolt HM, Bolt M, Kappus H (1977) Interaction of rifampicin treatment with pharmacokinetics and metabolism of ethinyloestradiol in man. Acta Endocr 85:189–197

Bonfils C, Dalet-Beluche I, Maurel P (1982) Induction by triacetyloleandomycin and partial purification of a LM3 form of cytochrome P-450 from rabbit liver microsomes. Biochem Biophys Res Commun 104:1011–1017

Bork RW, Muto T, Beaune PH, Srivastava PK, Lloyd RS, Guengerich FP (1989) Characterization of mRNA species related to human liver cytochrome P-450 nifedipine oxidase and the regulation of catalytic activity. J Biol Chem 264:910–919

Bourdi M, Larrey D, Nataf J, Berunau J, Pessayre D, Iwasaki M, Guengerich FP, Beaune PH (1990) A new anti-liver endoplasmic reticulum antibody directed against human cytochrome P-450 IA2: a specific marker of dihydralazine-induced hepatitis. J Clin Invest 85:1967–1973

Brian WR, Srivastava PK, Umbenhauer DR, Lloyd RS, Guengerich FP (1989) Expression of a human liver cytochrome P-450 protein with tolbutamide hydroxylase activity in Saccharomyces cerevisiae. Biochemistry 28:4993-4999

Brian WR, Sari M-A, Iwasaki M, Shimada T, Kaminsky LS, Guengerich FP (1990) Catalytic activities of human liver cytochrome P-450 IIIA4 expressed in Saccharomyces cerevisiae. Biochemistry 29:11280–11292

Brodie BB, Gillette JR, LaDu BN (1958) Enzymatic metabolism of drugs and other foreign compounds. Annu Rev Biochem 27:427–454

Buening MK, Fortner JG, Kappas A, Conney AH (1978) 7,8-Benzoflavone stimulates the metabolic activation of aflatoxin B_1 to mutagens by human liver. Biochem Biophys Res Commun 82:348–355

Buening MK, Chang RL, Huang MR, Fortner JG, Wood AW, Conney AH (1981) Activation and inhibition of benzo(a)pyrene and aflatoxin B_1 metabolism in human liver microsomes by naturally occurring flavonoids. Cancer Res 41:67–72

Butler MA, Iwasaki M, Guengerich FP, Kadlubar FF (1989) Human cytochrome P-450$_{PA}$ (P-450IA2), the phenacetin O-deethylase, is primarily responsible for the hepatic 3-demethylation of caffeine and N-oxidation of carcinogenic arylamines. Proc Natl Acad Sci USA 86:7696–7700

Chae YH, Yun CH, Guengerich FP, Kadlubar FF, El-Bayoumy K (1993) Roles of human hepatic and pulmonary cytochrome P450 enzymes in the metabolism of the environmental carcinogen 6-nitrochrysene. Cancer Res 53:2028–2034

Combalbert J, Fabre I, Fabre G, Dalet I, Derancourt J, Cano JP, Maurel P (1989) Metabolism of cyclosporin A. IV. Purification and identification of the rifampicin-inducible human liver cytochrome P-450 (cyclosporin A oxidase) as a product of P450IIIA gene subfamily. Drug Metab Dispos 17:197–207

Crespi CL, Penman BW, Steimel DT, Gelboin HV, Gonzalez FJ (1991) The development of a human cell line stably expressing human CYP3A4: role in the metabolic activation of aflatoxin B_1 and comparison to CYP1A2 and CYP2A3. Carcinogenesis 12:355–359

Cresteil T, Monsarrat B, Alvinerie P, Tréluyer JM, Vieira I, Wright M (1994) Taxol metabolism by human liver microsomes: identification of cytochrome P450 enzymes involved in its biotransformation. Cancer Res 54:386-392

Curi-Pedrosa R, Pichard L, Bonfils C, Jacqz-Aigrain E, Guengerich FP, Maurel P (1993) Major implication of cytochrome P450 3A4 in the oxidative metabolism of antisecretory drugs omeprazole and lansoprazole in human liver microsomes and hepatocytes. Abstracts, 5th European meeting of the International Society of the Study of Xenobiotics, Tours, 26–29 Sept, p 46

Elshourbagy NA, Guzelian PS (1980) Separation, purification, and characterization of a novel form of hepatic cytochrome P-450 from rats treated with pregnenolone-16α-carbonitrile. J Biol Chem 255:1279–1285

Ene MD, Roberts CJC (1987) Pharmacokinetics of nifedipine after oral administration in chronic liver disease. J Clin Pharmacol 27:1001–1004

Fisher CW, Shet MS, Martin-Wixtrom C, Estabrook RW (1992) High-level expression in Escherichia coli of enzymatically active fusion proteins containing the domains of cytochromes P450 and NADPH-P450 reductase flavoprotein. Proc Natl Acad Sci U S A 89:10817–10821

Fleming CM, Branch RA, Wilkinson GR, Guengerich FP (1992) Human liver microsomal N-hydroxylation of dapsone by cytochrome P-450 3A4. Mol Pharmacol 41:975–980

Ged C, Rouillon JM, Pichard L, Combalbert J, Bressot N, Bories P, Michel H, Beaune P, Maurel P (1989) The increase in urinary excretion of 6β-hydroxycortisol as a marker of human hepatic cytochrome P450IIIA induction. Br J Clin Pharmacol 28:373–387

Gillam EMJ, Baba T, Kim B-R, Ohmori S, Guengerich FP (1993) Expression of modified human cytochrome P450 3A4 in Escherichia coli and purification and reconstitution of the enzyme. Arch Biochem Biophys 305:123–131

Grace JM, Kinter MT, Macdonald TL (1994) Atypical metabolism of deprenyl and its enantiomer, (±)-N,α-dimethyl-N-propynylphenethylamine, by cytochrome P450 2D6. Chem Res Toxicol (in press)

Guengerich FP (1988) Oxidation of 17α-ethynylestradiol by human liver cytochrome P-450. Mol Pharmacol 33:500–508

Guengerich FP (1990) Mechanism-based inactivation of human liver cytochrome P-450 IIIA4 by gestodene. Chem Res Toxicol 3:363371

Guengerich FP, Kim D-H (1990) In vitro inhibition of dihydropyridine oxidation and aflatoxin B_1 activation in human liver microsomes by naringenin and other flavonoids. Carcinogenesis 11:2275–2279

Guengerich FP, Shimada T (1991) Oxidation of toxic and carcinogenic chemicals by human cytochrome P-450 enzymes. Chem Res Toxicol 4:391–407

Guengerich FP, Turvy CG (1991) Comparison of levels of several human microsomal cytochrome P-450 enzymes and epoxide hydrolase in normal and disease states using immunochemical analysis of surgical liver samples. J Pharmacol Exp Ther 256:1189–1194

Guengerich FP, Wang P, Davidson NK (1982) Estimation of isozymes of microsomal cytochrome P-450 in rats, rabbits, and humans using immunochemical staining coupled with sodium dodecyl sulfate-polyacrylamide gel electrophoresis. Biochemistry 21:1698–1706

Guengerich FP, Martin MV, Beaune PH, Kremers P, Wolff T, Waxman DJ (1986a) Characterization of rat and human liver microsomal cytochrome P-450 forms involved in nifedipine oxidation, a prototype for genetic polymorphism in oxidative drug metabolism. J Biol Chem 261:5051–5060

Guengerich FP, Müller-Enoch D, Blair IA (1986b) Oxidation of quinidine by human liver cytochrome P-450. Mol Pharmacol 30:287–295

Guengerich FP, Brian WR, Iwasaki M, Sari M-A, Bèèrnhielm C, Berntsson P (1991) Oxidation of dihydropyridine calcium channel blockers and analogues by human liver cytochrome P-450 IIIA4. J Med Chem 34:1838–1844

Guengerich FP, Kim BK, Gillam EMJ, Shimada T (1993) Mechanisms of enhancement and inhibition of cytochrome P450 catalytic activities. In: Lechner M (ed) 8th international conference on cytochrome P450: biochemistry, biophysics, and molecular biology, 24–28 Oct, Lisbon. Libbey Eurotext, Chichester

Halvorson M, Greenway D, Eberhart D, Fitzgerald K, Parkinson A (1990) Reconstitution of testosterone oxidation by purified rat cytochrome P450p (IIIA1). Arch Biochem Biophys 277:166–180

Harris JW, Rahman A, Kim B-R, Guengerich FP, Collins JM (1994) Metabolism of taxol by human hepatic microsomes and liver slices: participation of cytochrome P450 3A4 and of an unknown P450 enzyme. Cancer Res (submitted)

Howard PC, Aoyama T, Bauer SL, Gelboin HV, Gonzalez FJ (1990) The metabolism of 1-nitropyrene by human cytochromes P450. Carcinogenesis 11:1539–1542

Imaoka S, Enomoto K, Oda Y, Asada A, Fujimori M, Shimada T, Fujita S, Guengerich FP, Funae Y (1990) Lidocaine metabolism by human cytochrome P-450s purified from hepatic microsomes: comparison of those with rat hepatic cytochrome P-450s. J Pharmacol Exp Ther 255:1385–1391

Imaoka S, Imai Y, Shimada T, Funae Y (1992) Role of phospholipids in reconstituted cytochrome P450 3A forms and mechanism of their activation of catalytic activity. Biochemistry 31:6063–6069

Ingelman-Sundberg M, Johansson I, Hansson A (1979) Catalytic properties of the liver microsomal hydroxylase system in reconstituted phospholipid vesicles. Acta Biol Med Germ 38:379

Islam SA, Wolf CR, Lennard MS, Sternberg MJE (1991) A three-dimensional molecular template for substrates of human cytochrome P450 involved in debrisoquine 4-hydroxylation. Carcinogenesis 12:2211–2219

Itoh S, Yanagimoto T, Tagawa S, Hashimoto H, Kitamura R, Nakajima Y, Okochi T, Fujimoto S, Uchino J, Kamataki T (1992) Genomic organization of human fetal specific P-450IIIA7 (cytochrome P-450HFLa)-related gene(s) and interaction of transcriptional regulatory factor with its DNA element in the 5' flanking region. Biochim Biophys Acta 1130:133–138

Jaffe M (1879) Über die nach Einführung von Brombenzol und Chlorbenzol im Organismus entstehenden schwefelhaftigen Säuren. Ber Dtsch Chem Ges 12:1092

Jung-Hoffmann C, Kuhl H (1990) Pharmacokinetics and pharmacodynamics of oral contraceptive steroids: factors influencing steroid metabolism. Am J Obstet Gynecol 163:2183–2197

Keller W (1842) Über Verwandlung der Benzoesäure in Hippursäure. J Liebigs Ann Chem 43:108

Kitada M, Kamataki T (1979) Partial purification and properties of cytochrome P450 from homogenates of human fetal livers. Biochem Pharmacol 28:793–797

Kitada M, Kamataki T, Itahashi K, Rikihisa T, Kato R, Kanakubo Y (1985) Purification and properties of cytochrome P-450 from homogenates of human fetal livers. Arch Biochem Biophys 241:275–280

Kitada M, Kamataki T, Itahashi K, Rikihisa T, Kanakubo Y (1987) P-450 HFLa, a form of cytochrome P-450 purified from human fetal livers, is the 16α-hydroxylase of dehydroepiandrosterone 3-sulfate. J Biol Chem 262:13534–13537

Kitada M, Taneda M, Ohi H, Komori M, Itahashi K, Nagao M, Kamataki T (1989) Mutagenic activation of aflatoxin B₁ by P-450 HFLa in human fetal livers. Mutation Res 227:53–58

Kitada M, Taneda M, Ohta K, Nagashima K, Itahashi K, Kamataki T (1990) Metabolic activation of aflatoxin B₁ and 2-amino-3-methylimidazo[4, 5-f]-quinoline by human adult and fetal livers. Cancer Res 50:2641–2645

Kitamura R, Sato K, Sawada M, Itoh S, Kitada M, Komori M, Kamataki T (1992) Stable expression of cytochrome P450IIIA7 cDNA in human breast cancer cell line MCF-7 and its application to cytotoxicity testing. Arch Biochem Biophys 292:136–140

Kleinbloesem CH, van Brummelen P, Faber H, Danhof M, Vermeulen NPE, Breimer DD (1984) Variability in nifedipine pharmacokinetics and dynamics: a new oxidation polymorphism in man. Biochem Pharmacol 33:3721–3724

Kleinbloesem CH, van Harten J, Wilson JPH, Danhof M, van Brummelen P, Breimer DD (1986) Nifedipine: kinetics and hemodynamic effects in patients with liver cirrhosis after intravenous and oral administration. Clin Pharmacol Ther 40:21–28

Knodell RG, Browne D, Gwodz GP, Brian WR, Guengerich FP (1991) Differential inhibition of human liver cytochromes P-450 by cimetidine. Gastroenterology 101:1680–1691

Komori M, Nishio K, Ohi H, Kitada M, Kamataki T (1989) Molecular cloning and sequence analysis of cDNA containing entire coding region for human fetal liver cytochrome P-450. J Biochem 106:161–163

Komori M, Nishio K, Kitada M, Shiramatsu K, Muroya K, Soma M, Nagashima K, Kamataki T (1990) Fetus-specific expression of a form of cytochrome P-450 in human livers. Biochemistry 29:4430–4433

Koymans L, Vermeulen NPE, van Acker SABE, te Koppele JM, Heykants JJP, Lavrijsen K, Meuldermans W, Donné-Op den Kelder GM (1992) A predictive model for substrates of cytochrome P450-debrisoquine (2D6). Chem Res Toxicol 5:211–219

Krohn K, Uibo R, Aavik E, Peterson P, Savilahti K (1992) Identification by molecular cloning of an autoantigen associated with Addison's disease as steroid 17α-hydroxylase. Lancet 339:770–773

Kronbach T, Fischer V, Meyer UA (1988) Cyclosporine metabolism in human liver: identification of a cytochrome P-450III gene family as the major cyclosporine-metabolizing enzyme explains interactions of cyclosporine with other drugs. Clin Pharmacol Ther 43:630–635

Kronbach T, Mathys D, Umeno M, Gonzalez FJ, Meyer UA (1989) Oxidation of midazolam and triazolam by human liver cytochrome P450IIIA4. Mol Pharmacol 36:89–96

Lemoine A, Gautier JC, Azoulay D, Guengerich FP, Beaune P, Maurel P, Le-roux JP (1993) The major pathway of imipramine metabolism is catalyzed by cytochrome P-450 1A2 and P-450 3A4 in human liver. Mol Pharmacol 43:827–832

Lu AYH, Somogyi A, West S, Kuntzman R, Conney AH (1972) Pregneno-lone-16α-carbonitrile: a new type of inducer of drug-metabolizing enzymes. Arch Biochem Biophys 152:457–462

Macdonald TL, Gutheim WG, Martin RB, Guengerich FP (1989) Oxidation of substituted N,N-dimethylanilines by cytochrome P-450: estimation of the effective oxidation-reduction potential of cytochrome P-450. Biochemistry 28:2071–2077

McManus ME, Burgess WM, Veronese ME, Huggett A, Quattrochi LC, Tukey RH (1990) Metabolism of 2-acetylaminofluorene and benzo(a)pyrene and activation of food-derived heterocyclic amine mut-agens by human cytochromes P-450. Cancer Res 50:3367–3376

Miki N, Sugiyama T, Yamano T, Miyake Y (1981) Characterization of a highly purified form of cytochrome P-450 B_1. Biochem Int 3:217

Mimura M, Baba T, Yamazaki Y, Ohmori S, Inui Y, Gonzalez FJ, Guengerich FP, Shimada T (1993) Characterization of cytochrome P450 2B6 in human liver microsomes. Drug Metab Dispos 21:1048–1056

Miranda CL, Reed RL, Guengerich FP, Buhler DR (1991) Role of cytochrome P450IIIA4 in the metabolism of the pyrrolizidine alkaloid senecionine in human liver. Carcinogenesis 12:515–519

Molowa DT, Schuetz EG, Wrighton SA, Watkins PB, Kremers P, Mendez-Picon G, Parker GA, Guzelian PS (1986) Complete cDNA sequence of a cytochrome P-450 inducible by glucocorticoids in human liver. Proc Natl Acad Sci USA 83:5311–5315

Morel F, Beaune PH, Ratanasavanh D, Flinois J-P, Yang C-S, Guengerich FP, Guillouzo A (1990) Expression of cytochrome P-450 enzymes in cultured human hepatocytes. Eur J Biochem 191:437–444

Mueller GC, Miller JA (1953) The metabolism of methylated aminoazo dyes. II. Oxidative demethylation by rat liver homogenates. J Biol Chem 202:579–587

Oellerich M, Burdelski M, Lautz HU, Schulz M, Schmidt FW, Herrmann H (1990) Lidocaine metabolite formation as a measure of liver function in pa-tients with cirrhosis. TherDrug Monit 12:219226

Ohnhaus EE, Park BK (1979) Measurement of urinary 6β-hydroxycortisol ex-cretion as an in vivo parameter in the clinical assessment of the microsomal enzyme-inducing capacity of antipyrine, phenobarbitone and rifampicin. Eur J Clin Pharmacol 15:139–145

Park BK (1981) Assessment of urinary 6β-hydroxycortisol as an in vivo index of mixed function oxidase activity. Br J Clin Pharmacol 12:97–102

Patten C, Thomas PE, Guy R, Lee M, Gonzalez FJ, Guengerich FP, Yang CS (1993) Cytochrome P450 enzymes involved in acetaminophen activation by rat and human liver microsomes and their kinetics. Chem Res Toxicol 6:511–518

Pessayre D, Tinel M, Larrey D, Cobert B, Funck-Brentano C, Babany G (1983) Inactivation of cytochrome P-450 by a troleandomycin metabolite. Protective role of glutathione. J Pharmacol Exp Ther 224:685–691

Peyronneau MA, Renaud JP, Truan G, Urban P, Pompon D, Mansuy D (1992) Optimization of yeast-expressed human liver cytochrome-P450 3A4 catalytic activities by coexpressing NADPH-cytochrome P450 reductase and cytochrome b₅. Eur J Biochem 207:109–116

Raney KD, Shimada T, Kim D-H, Groopman JD, Harris TM, Guengerich FP (1992) Oxidation of aflatoxin B1 and related dihydrofurans by human liver microsomes: significance of aflatoxin Q1 as a detoxication product . Chem Res Toxicol 5:202–210

Regårdh CG, Edgar B, Olsson R, Kendall M, Collste P, Shansky C (1989) Pharmacokinetics of felodipine in patients with liver disease. Eur J Clin Pharmacol 36:473–479

Renaud J-P, Cullin C, Pompon D, Beaune P, Mansuy D (1990) Expression of human liver cytochrome P450 IIIA4 in yeast: a functional model for the hepatic enzyme. Eur J Biochem 194:889–896

Sattler M, Guengerich FP, Yun C-H, Christians U, Sewing K-F (1992) Human and rat liver microsomal cytochrome P450 3A enzymes are involved in biotransformation of FK506 and rapamycin. Drug Metab Dispos 20:753–761

Schellens JHM, Soons PA, Breimer DD (1988) Lack of bimodality in nifedipine plasma kinetics in a large population of healthy subjects. Biochem Pharmacol 37:2507–2510

Schuetz EG, Guzelian PS (1984) Induction of cytochrome P-450 by glucocorticoids in rat liver. II. Evidence that glucocorticoids regulate induction of cytochrome P-450 by a nonclassical receptor mechanism. J Biol Chem 259:2007–2012

Schuetz JD, Kauma S, Guzelian PS (1993) Identification of the fetal liver cytochrome CYP3A7 in human endometrium and placenta. J Clin Invest 92:1018–1024

Schwab GE, Raucy JL, Johnson EF (1988) Modulation of rabbit and human hepatic cytochrome P-450-catalyzed steroid hydroxylations by α-naphthoflavone. Mol Pharmacol 33:493–499

Shet MS, Fisher CW, Holmans PL, Arlotto MP, Estabrook (1993) Studies of drug and steroid metabolism using a purified recombinant artificial fusion protein containing human liver P450 3A4 and rat liver NADPH-P450 reductase (rF450[mHum3A4/mRatOR]L1). ISSX Proceedings 4:146 (5th North American meeting of the International Society of the Study of Xenobiotics, Tucson, 17–21 Oct)

Shimada T, Guengerich FP (1989) Evidence for cytochrome P-450$_{NF}$, the nifedipine oxidase, being the principal enzyme involved in the bioactivation of aflatoxins in human liver. Proc Natl Acad Sci U S A 86:462–465

Shimada T, Guengerich FP (1990) Inactivation of 1,3-, 1,6-, and 1,8-dinitropyrene by human and rat microsomes. Cancer Res 50:2036–2043

Shimada T, Iwasaki M, Martin MV, Guengerich FP (1989a) Human liver microsomal cytochrome P-450 enzymes involved in the bioactivation of procarcinogens detected by umu gene response in Salmonella typhimurium TA1535/pSK1002. Cancer Res 49:3218–3228

Shimada T, Martin MV, Pruess-Schwartz D, Marnett LJ, Guengerich FP (1989b) Roles of individual human cytochrome P-450 enzymes in the bioactivation of benzo(a)pyrene, 7,8-dihydroxy-7,8-dihydrobenzo(a)pyrene, and other dihydrodiol derivatives of polycyclic aromatic hydrocarbons. Cancer Res 49:6304–6312

Shimada T, Yamazaki H, Mimura M, Inui Y, Guengerich FP (1994) Interindividual variations in human liver cytochrome P450 enzymes involved in the oxidation of drugs, carcinogens, and toxic chemicals: studies with liver microsomes of 30 Japanese and 30 Caucasians. J Pharmacol Exp Ther (submitted)

Strobl GR, von Kruedener S, Stöckigt J, Guengerich FP, Wolff T (1993) Development of a pharmacophore for inhibition of human liver cytochrome P-450 2D6: molecular modeling and inhibition studies. J Med Chem 36: 1136–1145

Tateishi T, Krivorak Y, Wood AJJ, Guengerich FP, Wood M (1994) Identification of human liver P450 3A4 as the enzyme responsible for sulfentanil N-dealkylation. Abstracts, 12th international congress on pharmacology, 24–29 July (in press)

Thomas PE, Lu AYH, West SB, Ryan D, Miwa GT, Levin W (1977) Accessibility of cytochrome P450 in microsomal membranes: inhibition of metabolism by antibodies to cytochrome P450. Mol Pharmacol 13:819–831

Tinel M, Descatoire V, Larrey D, Loeper J, Labbe G, Letteron P, Pessayre D (1989) Effects of clarithromycin on cytochrome P-450. Comparison with other macrolides. J Pharmacol Exp Ther 250:746–751

Vanden Bossche H (1992) Inhibitors of P450-dependent steroid biosynthesis: from research to medical treatment. J Steroid Biochem Mol Biol 43:1003-1021

Vincent SH, Karanam BV, Painter SK, Chiu SHL (1992) In vitro metabolism of FK-506 in rat, rabbit, and human liver microsomes: identification of a major metabolite and of cytochrome P450 3A as the major enzymes responsible for its metabolism. Arch Biochem Biophys 294:454-460

Wang PP, Beaune P, Kaminsky LS, Dannan GA, Kadlubar FF, Larrey D, Guengerich FP (1983) Purification and characterization of six cytochrome P-450 isozymes from human liver microsomes. Biochemistry 22:5375-5383

Wang RW, Kari PH, Lu AYH, Thomas PE, Guengerich FP, Vyas KP (1991) Biotransformation of lovastatin. IV. Identification of cytochrome P-450 3A proteins as the major enzymes responsible for the oxidative metabolism of lovastatin in rat and human liver microsomes. Arch Biochem Biophys 290:355-361

Watkins PB, Wrighton SA, Maurel P, Schuetz EG, Mendez-Picon G, Parker GA, Guzelian PS (1985) Identification of an inducible form of cytochrome P-450 in human liver. Proc Natl Acad Sci USA 82:6310-6314

Watkins PB, Murray SA, Winkelman LG, Heuman DM, Wrighton SA, Guzelian PS (1989) Erythromycin breath test as an assay of glucocorticoid-inducible liver cytochrome P-450: studies in rats and patients. J Clin Invest 83:688-697

Watkins PB, Turgeon DK, Saenger P, Lown KS, Kolars JC, Hamilton T, Fishman K, Guzelian PS, Voorhees JJ (1992) Comparison of urinary 6-b-cortisol and the erythromycin breath test as measure of hepatic P450IIIA (CYP3A) activity. Clin Pharmacol Ther 52:265-273

Waxman DJ, Attisano C, Guengerich FP, Lapenson DP (1988) Cytochrome P-450 steroid hormone metabolism catalyzed by human liver microsomes. Arch Biochem Biophys 263:424-436

Waxman DJ, Lapenson DP, Aoyama T, Gelboin HV, Gonzalez FJ, Korzekwa K (1991) Steroid hormone hydroxylase specificities of eleven cDNA-expressed human cytochrome P450s. Arch Biochem Biophys 290:160166

Wrighton SA, Vandenbranden M (1989) Isolation and characterization of human fetal liver cytochrome P450HLp2: a third member of the P450III gene family. Arch Biochem Biophys 268:144-151

Wrighton SA, Ring BJ, Watkins PB, Vandenbranden M (1989) Identification of a polymorphically expressed member of the human cytochrome P-450III family. Mol Pharmacol 86:97-105

Wrighton SA, Brian WR, Sari MA, Iwasaki M, Guengerich FP, Raucy JL, Molowa DT, Vandenbranden M (1990) Studies on the expression and metabolic capabilities of human liver cytochrome P450IIIA5 (HLp3). Mol Pharmacol 38:207–213

Yamazaki H, Mimura M, Oda Y, Inui Y, Shiraga T, Iwasaki K, Guengerich FP, Shimada T (1993) Roles of different forms of cytochrome P450 in the activation of the promutagen 6-aminochrysene to genotoxic metabolites in human liver microsomes. Carcinogenesis 14:1271–1278

Yun C-H, Shimada T, Guengerich FP (1992a) Contributions of human liver cytochrome P-450 enzymes to the N-oxidation of 4,4'-methylene-bis(2-chloroaniline). Carcinogenesis 13:217–222

Yun C-H, Shimada T, Guengerich FP (1992b) Roles of human liver cytochrome P-4502C and 3A enzymes in the 3-hydroxylation of benzo(a)pyrene. Cancer Res 52:1868–1874

Yun C-H, Wood M, Wood AJJ, Guengerich FP (1992c) Identification of the pharmacogenetic determinants of alfentanil metabolism: cytochrome P-450 3A4. An explanation of the variable elimination clearance. Anesthesiology 77:467–474

Yun C-H, Okerholm RA, Guengerich FP (1993) Oxidation of the antihistaminic drug terfenadine in human liver microsomes: role of cytochrome P450 3A(4) in N-dealkylation and C-hydroxylation. Drug Metab Dispos 21:403409

Zanger UM, Hauri HP, Loeper J, Homberg JC, Meyer UA (1988) Antibodies against human cytochrome P-450dbl in autoimmune hepatitis type II. Proc Natl Acad Sci USA 85:8256–2860

Zilly W, Breimer DD, Richter E (1975) Induction of drug metabolism in man after rifampicin treatment measured by increased hexobarbital and tolbutamide clearance. Eur J Clin Pharmacol 9:219–227

10 Strategies for Verifying the Involvement of Specific Cytochrome P450 Enzymes in Xenobiotic Metabolism

P. H. Beaune, S. Lecoeur, C. Belloc, A. Lemoine, D. Pompon,
J.-C. Gautier, and H. Kroemer

10.1   Introduction 187
10.2   Material and Methods 188
10.3   Results and Discussion 188
References 192

10.1 Introduction

Xenobiotics are very often hydrophobic compounds and therefore cannot be eliminated as such from the human body but must be metabolized by xenobiotic metabolizing enzymes (XME) to be excreted in bile or urines. These enzymes are usually classified as phase I or phase II (Williams 1967), and among the most important phase I enzymes are cytochromes P450. Many P450 enzymes have been purified, cloned, and classified according to an internationally accepted nomenclature (Nelson et al. 1993). Each P450 has an overlapping substrate specificity and may metabolize a xenobiotic into a more or less active or a more or less toxic compound. The amount of each P450 is variable depending on genetic and environmental factors (Gonzalez 1990). It is therefore very important to know the role and the concentration of each P450 in the human organism in order to understand and to foresee the pharmacological and toxicological potential of a given xenobiotic in a given individual. Heterologous systems allowing the expression of cloned P450 are new tools that are very useful in the understanding of xenobiotic metabolism since they allow working with a unique pure enzyme. This paper reviews the way in which these systems can be used for this purpose and be integrated into a general strategy for answering the question of which P450 produces which metabolite from a given xenobiotic. It is also shown how these systems were used to decipher the first steps of triggering tienilic acid induced hepatitis (Beaune et al. 1987; Lecoeur et al. 1994; Lopez-Garcia et al 1993).

10.2 Material and Methods

Human cytochromes P450 1A1, 1A2, 2C8, 2C9, 2C18, 2D6, 2E1, and 3A4 were expressed in yeasts as described previously (Gautier et al. 1993; Cosme et al., unpublished data; Lecoeur et al. 1994; Baird et al., unpublished data; Renaud et al. 1991). The yeast strains overexpressing NADPH cytochrome P450 reductase has been described by Truan et al. (1993). Yeast or human liver microsomes were prepared as previously described (Renaud et al. 1991) P450 1A1, 1A2, 2C9, 2E1, and 3A4 were also expressed in bacteria. Two expression plasmids were used, pET and pGex (Novagen, Switzerland, Pharmacia Sweden). These were used according to the manufacturer's recommendations, and inactive proteins or fragments of proteins were produced in *Escherichia coli* after induction by isopropylthiogalactoside. P 450 2D6 cDNA was a generous gift from Dr. U. Meyer, and the other P450 cDNAs were obtained by polymerase chain reaction amplification of human liver mRNAs. After insertion in pUC19, the cDNAs were sequenced, and very few mutations were observed when compared to published sequences. Peptides expressed in *E. coli* were used as antigen either in immunoblots or for antibodies production.

10.3 Results and Discussion

A general strategy can be defined for determining which P450 produces which metabolite(s) from a given xenobiotic. We concentrate essentially on human liver.

It is first necessary to measure the production of the metabolite and to have a set of human liver microsomes for assaying the activity in different samples containing various amounts of P450s. Then concentration of the individual's P450 can then be measured by immunoblots in the same human liver microsome bank. This necessitates antibodies produced against purified P450s; the specificity of the antibodies depends on the purity of the injected antigen and on the cross-reactivity with other P450s. To increase the specificity we expressed and purified cloned P450s or fragments of P450s in bacteria and thus obtained antibodies against P450 1A1, 1A2, 2C9, 2E1, and 3A4, which are quite specific even, in some cases, inside a P450 subfamily. The amount of P450 measured by immunoblot must then be correlated with the rate of metabolite production. Such a correlation indicates that a P450 may be responsible for the production of the metabolite, but this must be confirmed by other means.

The second approach is immunoinhibition. Specific and inhibitory antibodies are necessary. Up to now the antibodies that we have produced against peptides (fragment of P450) expressed in bacteria have not been inhibitory except anti-P4503A4 and 2E1. This may be because they were produced as inclusion bodies. The reliability of this approach depends on the specificity of the antibody.

The third complementary approach is the use of specific inhibitors. This is more accessible because it does not require the use of antibodies, and the chemical inhibitors are commercially available; however, several conditions must be filled: the specificity of the inhibitor must be known and controlled, and the inhibition must be of competitive type. The P450s expressed in heterologous systems are very useful in determining the specificity of the known inhibitors and in characterizing new ones (Guengerich 1994).

Some P450 are inducible, and this property can be used to help determine the P450 responsible for the production of a metabolite. Indeed, in human hepatocytes, P450 3A can be induced by rifampicine (Morel et al. 1990), P450 1A by omeprazole or aromatic hydrocarbons (Diaz et al. 1990), and P450 2E1 by ethanol. If the production of a metabolite follows the increase in a specific P450, it also indicates that this P450 is implicated in this production.

Finally, it is important to show that a pure P450 can produce the studied metabolite. Purified P450 in a reconstituted system was the first solution, but it has at least two drawbacks: (a) the purification of human P450s is tedious and difficult, and moreover the availability of human liver is constantly decreasing; (b) the purity of the P450 is never certain. Nevertheless the first informations on human P450 specificity were brought by this approach (review in Guengerich 1989). The cloning of human P450 cDNAs and the progress in heterologous expression systems has allowed determination of substrate specificity of several human P450s. However the results obtained must be considered in the light of the fact that only one P450 is expressed in these systems while in human liver several P450s are expressed at different levels; it is thus crucial before drawing conclusions on the importance of a P450 in the production of a metabolite to consider the affinity of the enzyme for the substrate and the concentration of the P450 in human liver (Guengerich and Turvy 1991). Several examples illustrate this strategy:

Imipramine is an important drug which is metabolized according to three main pathways: one leading to the 2-hydroxy imipramine which depends on P450 2D6, one leading to 10-hydroxy imipramine, and one leading to the N-demethylated product. The enzymes producing the two last metabolites were unknown, and by using a combination of all the above means we showed that the N-demethylation depends on both P450 12 and 3A4, the latter being the most important. This approach was validated by the fact that we confirmed that P450 2D6 was essentially producing the 2-hydroxy metabolite, which was consistent with both in vitro and in vivo results (Brosen et al. 1991). One of the important points was that we compared the results obtained by the different approaches, and that they were consistent with each other. For instance, we measured or calculated the production of the three main metabolites in the three systems used in this paper: human liver microsomes, human hepatocytes, and yeast microsomes. Table 1 shows clearly that the results obtained were quite comparable, regardless of which system was used. Similar results have been obtained with propafenone and verapamil (Botsch et al. 1993; Kroemer et al. 1993). It was shown that the N-dealkylpropafenone is produced essentially by P450 3A4 and 1A2. It was also shown that P450 1A1 has the highest turnover number for propafenone N-dealkylation; nevertheless, due to its low level of expression in human liver, its actual participation in the production of N-dealkylpropafenone is negligible. This lack of real implication of P450 1A1 in the metabolism could be changed by induc-

Table 1. Comparison between yeast, human liver microsomes and human hepatocytes

Metabolite	Human hepatocytes	Human liver microsomes	Yeasts
2-OH-imipramine	32	10	26
10-OH-imipramine	8	6	11
Desipramine	60	84	63

In the three systems the percentage of each metabolites was calculated with the data from the paper of Lemoine et al. (1993). For yeast the production of each metabolite was calculated by multiplying the turnover number by the estimated level of each individual P450 (Guengerich and Turvy 1991). P450 1A1 was considered under 5 pmol/mg protein which should be roughly the limit of detection of Western blots. Yeasts expressed P450 were: 1A2, 1A2, 2C9, 2D6, and 3A4

tion. For instance, it has been shown that chlorzoxazone 6-hydroxylation is catalyzed by P450 2E1 (Peter et al. 1990); more recently Carrière et al. (1993) confirmed that P450 1A1 expressed in yeast is also able to catalyze this 6 hydroxylation; in addition, they showed that treatment of human hepatocytes in primary culture by 3-methylcholanthrene increased the chlorzoxazone 6-hydroxylation although P450 2E1 was not increased. The augmentation of this catalytic activity was essentially due to the P450 1A1 induction since P450 1A2 is unable to metabolize chlorzoxazone.

Finally, it was shown that verapamil metabolism is essentially due to P450 3A4 and 1A2 (Kroemer et al. 1993). These two P450s seem in general to be very important in xenobiotic metabolism, either by their higher turnover numbers (P450 1A2) or by their high level of expression in human liver (P450 3A4; Guengerich and Turvy 1991).

We have also applied this heterologous expression technology to the immunotoxicology field. Tienilic acid is a drug which was marketed in the United States and France as an uricosuric diuretic. This drug triggered, in very rare occasions, an immunoallergic type hepatitis characterized by the presence of autoantibodies (anti-LKM2) in the serum of the patients (Beaune et al. 1987). The scheme of Fig. 1 was postulated to explain the first steps of the disease:

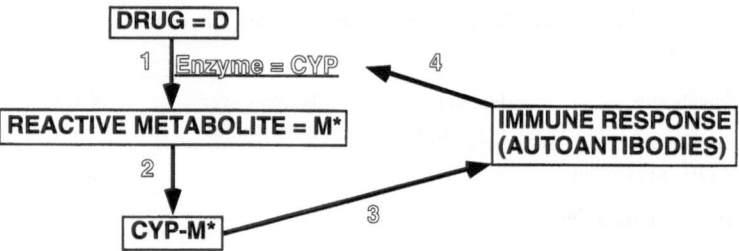

Fig.1. Postulated scheme of the first steps of a drug induced hepatitis of immunoallergic type

Yeasts expressing a single P450 were used to confirm some of these steps. P450 2C9 was shown to be the P450 responsible of the production of a reactive metabolite; all other tested P450s were unable to produce significant amounts of this reactive metabolite covalently bound to proteins (Dansette et al. 1991; Lopez-Garcia et al. 1993; Lecoeur et al. 1994). It was also shown that P450 2C9 is the main P450 on which the reactive metabolite is bound (Lecoeur et al. 1994). Finally, P450 2C9 is the only target of the autoantibodies which does not recognize any other P450 expressed in yeast, even the P450s of the same subfamily, namely P460 2C8 and 2C18. Therefore most of the postulated steps were clearly verified thanks to P450 expressed in yeasts or bacteria.

In conclusion, this new technology has allowed substantial progress in the field of xenobiotic metabolism and toxicity, but it must be used in conjunction with other methods to strengthen the results obtained.

Acknowledgements. Part of the results presented here were obtained the support of the Ministère de la recherche and Rhône-Poulenc-Rorer through the Bioavenir program.

References

Beaune PH, Dansette PM, Mansuy D, Kiffel J, Finck M, Amar C, Leroux JP, Homberg JC (1987) Human anti-endoplasmic reticulum autoantibodies appearing in a drug-induced hepatitis are directed against a human liver cytochrome P450 that hydroxylates the drug. Proc Natl Acad Sci USA 84:551–555

Botsch S, Gautier JC, Beaune PH, Eichelbaum M, Kroemer H (1993) Identification and characterztion of the cytochrome P450 enzymes involved in N-
dealkylation of propafenone: molecular base for interaction potential and
variable disposition of active metabolites. Mol Pharmacol 43:120–126

Brosen K, Zeugin T, Meyer UA (1991) Role of P4502D6, the target of the
spartein-debrisoquine oxidation polymorphism in the metabolism of imipramine. Clin Pharmacol Ther 49:18–23

Carrière V, Goasduff T, Ratanasananh D, Morel, F, Gautier JC, Guillouzo A,
Beaune PH, Berthou F (1993) Both cytochrome P4502E1 and 1A1 are involved in the metabolism of chlorzoxazone. Chem Res Toxicol 6:852–857

Dansette PM, Amar C, Valadon P, Pons C, Beaune PH, Mansuy D (1991) Hydroxylation and formation of electrophilic metabolites of tienilic acid and
its isomer by human liver microsomes: catalysis by a P450 2C different
from that responsible for mephenytoin hydroxylation. Biochem Pharmacol
41:553–560

Diaz D, Fabre I, Daujat M, Saint-Aubert B, Bories P, Michel H, Maurel P
(1990) Omeprazole is an aryl hydrocarbon-like inducer of human hepatic
cytochrome P450. Gastroenterology 99:737–747

Gautier JC, Urban P, Beaune PH, Pompon D (1993) Engineered yeast cells as
model to study coupling between human xenobiotic metabolizing enzymes.
Eur J Biochem 211:63–72

Gonzalez FJ (1990) Molecular genetics of the P450 superfamily. Pharmacol
Ther 45:139

Guengerich FP (1989) Characterization of human microsomal cytochromes
P450. Annu Rev Pharmacol Toxicol 29:241–264

Guengerich FP (1994) Cytochromes P450 of human liver. Classification and
activity profiles of the major enzymes. In: Pacicici GN (ed) Advances in
drug metabolism (in press)

Guengerich FP, Turvy CG (1991) Comparison of levels of several human liver
microsomal cytochrome P450 enzymes and epoxide hydrolase in normal
and disease states using immunochemical analysis of surgical liver
samples. J Pharmacol Exp Ther 256:1189–1194

Kroemer H, Gautier JC, BeaunePH, Henderson C, Wolf CR, Eichelbaum M
(1993) Identification of P450 enzymes involved in metabolism of Verapamil in humans. Naunyn-Schmiedbergs Arch Pharmacol 348:332–337

Lecoeur S, Bonierbale E, Challine D, Gautier JC, Valadon P, Dansette PM,
Catinot R, Ballet F, Mansuy D, Beaune PH (1994) Specificity of in vitro
covalent binding of tienilic acid metabolites to human liver microsomes in
relationship to the type of hepatotoxicity. Chem Res Toxicol (in press)

Lopez-Garcia MP, Dansette PM, Valadon P, Amar C, Beaune PH, Guengerich
FP, Mansuy D (1993) Human liver P450s expressed in yeast as tools for

reactive metabolite formation studies: oxidative activation of tienilic acid by P4502C9. Eur J Biochem 213:223–232

Morel F, Beaune PH, Ratanasananh D, Flinois JP, Yang CS, Guengerich, FP, Guillouzo A (1990) Expression of cytochrome P450 enzymes in cultured human hepatocytes. Eur J Biochem 191:437–444

Nelson DR, Kamataki T, Waxman DJ, Guengerich FP, Estabrook RW, Feyersen R, Gonzalez FJ, Coon MJ, Gunsalus IC, Gotoh O, Okuda K, Nebert DW (1993) The P450 superfamily: update on new sequences, gene mapping, accession numbers, early trivial names of enzymes and nomenclature. DNA Cell Biol 12:1–51

Peter R, Böcker R, Beaune PH, Iwasaki M, Guengerich FP, Yang CS (1990) Hydroxylation of chlorzoxazone as a specific probe for human liver cytochrome P450 2E1. Chem Res Toxicol 3:566–573

Renaud JP, Cullin C, Pompon D, Beaune PH, Mansuy D (1991) Expression of human liver P450 3A4 in yeast. Eur J Biochem 194:889–896

Truan G, Cullin C, Reisdorf P, Urban P, Pompon D (1993) Enhanced in vivo monooxygenase activities of mammalian P450s in engineered yeasts cells producing high levels of NADPH P450 reductase and human cytochrome b5. Gene 125:49–55

Williams RT (1967) The biogenesis of conjugation and detoxication products. In: Bernfeld P (ed) Biogenesis of natural products. Pergamon, New York, pp 589-639

11 Regulation of Cytochromes P450 by Substrate Interactions

M. Ingelman-Sundberg, A. Zhukov, S. Mkrtchian, and E. Eliasson

11.1   Introduction 195
11.2   Substrate Regulation of Cytochrome P450 Redox Cycling 197
11.3   Substrate-Dependent Regulation of P450 Phosphorylation
        and Degradation. 198
11.4   Substrate-Governed Transport of Cytochrome P450
        to Lysosomes 200
11.5   Substrate-Regulated Degradation of Cytochrome P450
        in the Endoplasmic Reticulum. 201
11.6   Proteolytic Systems in the Endoplasmic Reticulum 203
11.7   Alternative Pathways for Rapid P450 Degradation 204
11.8   Conclusions 206
References 206

11.1 Introduction

The level of several hepatic cytochrome P450 forms is to a great extent controlled by the apparent concentration of substrates in the liver. This is of course beneficial for the control of the cellular levels of both endogenous and exogenous compounds. It is probable that this control system has evolved for survival purposes of the organisms, but an endogenously driven evolution in some specific cases cannot be excluded. The substrate-dependent P450 regulation is achieved at the transcriptional, pretranslational, translational, and posttranslational levels. Transcriptional activation of the P450 gene is by far the most

Table 1. Half-lives of rat hepatic P450s: mechanisms for induction (from Watkins et al. 1986 and references therein; Song et al. 1989; Eliasson et al. 1990, 1994)

Isozyme	$t_{1/2_1}$	$t_{1/2_2}$	Transcriptional	Posttranslational
CYP1A1	–	37 h	Yes	No
CYP2B1	–	37 h	Yes	No
CYP2E1	7 h	37 h	Yes	Yes
CYP3A1	14 h	63 h	Yes	Yes

Table 2. Levels for substrate- and hormonal-dependent regulation of CYP2E1

Level	Agent/conditions	Extent	References
Transcription	Starvation	3- to 4-fold	Johansson et al. 1990
	Chronic ethanol	3- to 4-fold	Badger et al. 1993
mRNA stabilization	Diabetes	3-fold	Song et al. 1987
Translation efficiency	Isoniazide	2-fold	Park et al. 1993
Enzyme stabilization	Ethanol	3- to 4-fold	Eliasson et al. 1988
	Imidazole		Eliasson et al. 1988
	Isoniazide		Eliasson et al. 1990
	Acetone		Song et al. 1989
	Maximum	54-fold	

common mechanism (see Porter and Coon 1991, for review). However, certain isozymes with biphasic turnover are regulated by substrate both at the transcriptional and posttranslational levels (Table 1). Among the alternative cellular mechanisms of raising the enzyme level, the post-translational mechanism represents a very rapid and efficient way, provided that the protein synthesis continues.

With CYP2E1 as an example, it is evident that the level of the enzyme in rat can fluctuate at least 20-fold, depending on the diet in combination with ethanol (Ronis et al. 1993). The various substrate- and hormonal-dependent mechanisms described to regulate this isozyme are summarized in Table 2. As described for other enzymes with important endogenous functions, such as 3-hydroxy-3-methylgluaryl coenzyme A reductase (Goldstein and Brown 1990), the cell utilizes the concerted action of several different mechanisms to raise the level of the enzyme in a pronounced manner. Thus, the maximum capability for induction of CYP2E1 is about 54-fold (Table 2).

This principle is well illustrated for ethanol-dependent regulation of CYP2E1 in the total enteral nutrition model, where ethanol is pumped intragastrically in a continuous fashion. This kind of ethanol treatment causes the blood (BAC) and urinary alcohol (UAC) levels to fluctuate in a cyclic manner (Badger et al. 1993) and CYP2E1 levels follow in a concomitant and a proportional manner up to at least BAC = 400 mg/dl. At BAC below 200 mg/dl this increase is entirely caused by posttranslational stabilization of CYP2E1, as evident from no increase in mRNA, whereas at BAC above 200 mg/dl transcriptional gene activation is seen (Badger et al. 1993; Ronis et al. 1993). Thus far, the manner by which ethanol affects the increased gene transcription is unknown, but it is probably related to indirect effects via hormonal changes.

11.2 Substrate Regulation of Cytochrome P450 Redox Cycling

The actual step of substrate oxygenation by cytochrome P450 is preceded by several steps of electron transfer and oxygen activation during which the enzyme-bound substrate remains chemically intact. Still, the substrate affects these preliminary steps, most commonly by acceleration of electron transfer to cytochrome P450 due to substrate-induced low-to-high spin transition of the heme iron. Substrate control of electron flow is especially strong in bacteria, where the rate of electron transfer from NADH to oxygen is negligible in the absence of substrate, an obvious regulatory advantage preventing the waste of reducing equivalents (Sligar and Murray 1986). In higher organisms such control is substantially weaker, and electron transfer from NADPH to O_2 can still take place in the absence of substrate resulting in the formation of O_2^-, H_2O_2, and water via 1-, 2-, and 4-electron reduction of O_2 (Zhukov and Archakov 1982; Gorsky et al. 1984; Blanck et al. 1991; Hanukoglu et al. 1993a). Substrate can suppress the oxidase activities as is the case with cytochrome P450scc and cholesterol (Hanukoglu et al. 1993a), where the substrate appears to act by stabilizing oxycytochrome P450 (Tuckey and Kamin 1982). The substrate can also activate oxidase reactions, which is observed with fluorinated hydrocarbons which divert electron flow mainly towards 4-electron oxygen reduction to water (Gorsky et al. 1984; Zhukov and Archakov 1985; Wang et al. 1993). Superoxide radicals, H_2O_2, and especially hydroxyl radicals can modify P450 heme and apoprotein leading to enzyme inactivation (see Karuzina and Archakov 1994, for review).

Hanukoglu et al. (1993b) point out that cytochrome P450c11, which shows greater than 50% uncoupling during the metabolism of deoxycorticosterone, undergoes rapid steroid-dependent inactivation in steroidogenic cells, while cytochrome P450scc characterized by less than 15% leakage in the presence of cholesterol is much more stable. The destabilizing effect of different steroids on P450c11 in cells is correlated with their ability to induce the leakage of the mono-oxygenase cycle vitro (Hanukoglu et al. 1993b). Several papers by Stadtman et al. indicate the importance of oxy-modification of enzymes as a trigger for degradation (see Karuzina and Archakov 1994). Thus, substrate control of cytochrome P450 oxidase activity can represent another way of controlling cytochrome P450 level in the cell, although its significance as a physiological regulatory mechanism is still unknown.

11.3 Substrate-Dependent Regulation of P450 Phosphorylation and Degradation

The mechanism for a substrate-dependent action on the CYP2E1 level can well be studied in primary hepatocytes, which upon isolation lose their ability to synthesize CYP2E1 (Eliasson et al. 1988). In this system we found that substrates for CYP2E1 can prevent degradation of the protein at an efficiency similar in magnitude to their binding affinities, as determined by spectrophotometric spin-shift analysis upon substrate-binding. By contrast, the substrates were not effective in preventing CYP2B1 degradation in the same hepatocytes, isolated from starved and acetone-treated rats, where indeed CYP2B1 together with CYP2E1 represent the major hepatic P450 isozymes (Johansson et al. 1988).

Obviously, substrates interacted with an endogenous system controlling the rate of P450 degradation. Screening for this system was carried out by investigating the effects of different hormones on P450 degradation in isolated hepatocytes, in the presence or absence of substrates. It turned out that cAMP-analogues or adenylyl cyclase stimulating hormones, such as glucagon or adrenaline, increase the rate of CYP2E1 degradation without affecting CYP2B1 disappearance (Eliasson et al. 1990, Johansson et al. 1991). Substrates for CYP2E1 prevent the hormonal effect, which indicates that by binding to P450 they changed the conformation of the enzyme preventing it from interaction with the degradation system.

It has been found that all agents that increased CYP2E1 degradation also causes phosphorylation of the enzyme on Ser-129, and a correlation has been observed between the extent of enzyme phosphorylation and degradation ($r = 0.93$–0.96). All substrates tested prevented this phosphorylation, as well as degradation (Eliasson et al. 1990). By contrast, CYP2B1 degradation in the hepatocytes was unaffected by the hormones, despite being phosphorylated at the corresponding site, Ser-128 (Pyerin and Taniguchi 1989). The structural explanation for this differential behavior of the two isozymes is unclear, despite their 55% amino acid homology.

The phosphorylation of Ser-129 apparently triggered a rapid heme loss (Eliasson et al. 1990). The region around the phosphorylatable Ser-129 is very well conserved among P450s belonging to gene family 2. Extrapolation of the three-dimensional structure of CYP101 to the mammalian microsomal P450 forms suggests that the region around Ser-129 is involved in substrate recognition and related to the heme pocket, apparently explaining the sensitivity for heme loss after phosphorylation (Gotoh 1992; Poulos et al. 1985).

A similar mechanism for substrate-regulated phosphorylation of P450 has also been found regarding glucocorticoid CYP3A1. This isozyme has previously been reported to be substrate-stabilized in vivo and in isolated hepatocytes by macrolide antibiotics (see Watkins et al. 1986). It has also been shown that clotrimazole (CTZ), an antimycotic agent, described as a most potent inducer of CYP3A1 in vivo, acts mainly by causing posttranslational accumulation of the CYP3A1 protein (Hostetler et al. 1989). The primary structure of CYP3A1 has only

small parts with homology to CYP2E1, and the well-conserved phosphorylatable sequence around Ser-129 is absent.

In hepatocytes isolated from dexamethasone-treated rats, we found
that CYP3A1 was efficiently phosphorylated after stimulation of the
cells with either cAMP or glucagon, and that the reaction was inhibited
in the presence of CTZ. Data from in vitro experiments indicate that
phosphorylation causes rapid denaturation of the enzyme. The site phosphorylated was evaluated by protease digestion of the phosphorylated
enzyme using several different proteases, followed by HPLC analysis of
radioactive peptides and radiosequencing (Eliasson et al. 1994). The
phosphorylation site thus identified was Ser-393. P450 sequence alignment analysis (Nelson and Strobel 1988), including comparisons with
the characterized three-dimensional structure of CYP101, suggests that
Ser-393 is located in a flexible part of the P450 molecule, poor in
α-helixes, but rich in β-structures. The corresponding CYP101 residue
Asp-316 is included in the β-3 structure described to interact with the
heme edge, and also to restrict the substrate-binding pocket (Poulos et
al. 1985).

We also observed that phosphorylation causing denaturation of
CYP3A1 caused rapid aggregation of CYP3A1, causing a selective loss
of this band from the sodium dodecyl sulfate–polyacrimed gel electrophoresis gels. Sodium dodecyl sulfate resistant, high Mr aggregates
were detected in western blotting. From this we suggest that phosphory-
lation-dependent aggregation may be an event preceding proteolysis of
this P450.

11.4 Substrate-Governed Transport of Cytochrome P450 to Lysosomes

As discussed above, P450 substrate binding is a crucial determinant for
the intracellular fate of certain P450 forms. It turns out that not only the
stability of a P450 but even the intracellular distribution is affected by
substrate binding. An example of this concerns the intracellular locus of
proteolysis, where differences apparently exist between different P450
isozymes. Thus, CYP2B1 having only a monophasic degradation with a
longer half-life (Table 1) is degraded apparently only according to the
autophagosomal-lysosomal pathway (Masaki et al.1987; Ronis et al.

1991a). In the case of CYP2E1 it is evident that the substrate can shift the degradation pattern from a biphasic to a monophasic degradation with long half-time, about 37 h (Song et al. 1989). The relatively rapid turnover of CYP2E1 in the absence of substrate and the efficient protection by substrate of this phase, causes a relatively rapid increase of the isozyme in vivo after a single injection of a substrate, as exemplified with acetone (Ronis et al. 1991). By contrast, acetone-dependent induction of CYP2B1 proceeded after a lag phase. Using electron microscopy and immunoblot analysis, it was shown that the slower phase for degradation has its background in a pathway of degradation according to the autophagosomal-lysosomal pathway (Ronis et al. 1991). Thus, in the presence of substrate CYP2E1 was associated with autophagosomes and lysosomes. After injection of acetone to the rats the amount of CYP2E1 in the lysosomes increased threefold (Ronis et al. 1991). This indicates that substrate binding mediates an increase of CYP2E1 transported from the endoplasmic reticulum (ER) to lysosomes. This could suggest that rapid degradation in the absence of substrate takes place at another subcellular location, perhaps in the ER itself.

11.5 Substrate-Regulated Degradation of Cytochrome P450 in the Endoplasmic Reticulum

The function of the ER is synthesis, control of correct folding and processing of proteins, in particular of those which are exported to the lysosomes and out of the cell. In the synthesis machinery the folding process appears to be a critical event, and incorrectly folded protein products are recognized by proteolytic systems in the ER and degraded within the compartment. Accumulation of unfolded proteins can be sensed by an ER kinase (Ire 1 p) which signals to the nucleus for increased protein synthesis (see Shamu et al. 1994).

ER degradation has been shown to be the case for many proteins by now (Table 3), among them the subunits of the T-cell receptor (Klausner and Sitia 1990), a subunit of the asialoglycoprotein receptor if it failed to assemble correctly (Amara et al. 1989) and unassembled immunoglobulin light-chain produced in CH12 lymphoma cells.

The incorrectly assembled proteins never reach the Golgi apparatus, and inhibitors of secretorial pathways as well as inhibitors of lysosomal

Table 3. Candidates for degradation in the endoplasmic reticulum (Gardner et al. 1993)

Murine T-cell antigen receptor subunits α, β, χ	Lippincott-Schwartz et al. 1988 Bonifacino et al. 1991
Human T-cell antigen receptor subunits α, β, δ	Wileman et al. 1991
Asialoglycoprotein receptor subunit H2a,b	Wikström et al. 1992
Acetylcholine receptor subunits	Claudio et al. 1989
Mutant transferrin receptor	Hoe and Hunt 1992
HL-DRβ	Cotner 1992
IgM	Sitia et al. 1987
Apolipoprotein B	Sato et al. 1990
Truncated influenza hemagglutinin	Doyle et al. 1986
Mutant low density lipoprotein receptor	Esser and Rusell 1988
Truncated β-hexoaminidase	Lau and Neufeld 1989
α_1-Antitrypsin PiZ variant	Le et al. 1990, 1992
Chimeric globin	Stoller and Shields 1989
CD18	Kishimoto et al. 1987
Acetylcholine esterase	Rotundo 1988
3-Hydroxy-3-methylglutaryl CoA reductase	Chun et al. 1990
Cystic fibrosis TM conductance regulator	Cheng et al. 1990
CYP2E1	Eliasson et al. 1992

function do not influence their degradation. The protein quality control could take place even in compartments associated with the ER. Thus, carboxy-terminally truncated rebophorin 1 has characteristics for a degradation event, indicating that a second pre-Golgi compartment, probably the so called intermediate compartment, is involved in degradation of proteins in the secretory pathway (Tsao et al. 1992). The ER degradation event is not limited to membrane proteins, and soluble proteins also have the same degradation characteristics. Examples of those are mutated forms of secretory proteins, for example, variants of α_1 antitrypsin, (Le et al. 1990). Another ER compartment proposed to be associated with mitochondria has been described as the location for proteolytic cleavage of haptoglobin (Wassler and Fries 1993). This particular subcompartment of ER sediments together with mitochondria, and the responsible protease is active at a neutral or slightly alkaline pH. The degradation was GTP-dependent, probably by its action as a membrane fusion agent. The pre-Golgi degradation ER-associated proteolysis generally characterized by its insensitivity to lysosomotrophic agents, such as NH_4Cl or chloroquine.

Degradation of CYP2E1 in hepatocytes was found to follow these characteristics of ER degradation (Eliasson et al. 1992) and the stimulatory effect of MgATP on the disappearance of the enzyme in microsomes under conditions that favored phosphorylation was taken as an indication for the presence of a microsomal proteolytic system active in degradation of P450s (Eliasson et al. 1992). We have therefore put effort in the identification and isolation of such a system.

11.6 Proteolytic Systems in the Endoplasmic Reticulum

Several serine, cystein, and metallo proteinases have been identified in microsomes thus far. Some of them are believed to be responsible for preprotein processing (Brennan and Peach 1988, Kawabata and Davie 1992) or cleavage of signal peptides (Evans et al. 1986; Baker and Lively 1987) while the functions of others are unknown (Urade et al. 1992; Tamanoe et al. 1993). Based upon the assumption that a protease localized within the endoplasmic reticulum and active in cytochrome P450 degradation recognizes mainly forms of the enzyme being phosphorylated and unfolded in nature, we purified two serine proteases from rat liver microsomes which are active in CYP2E1 degradation (Zhukov et al. 1993). An assay system was employed in which the CYP2E1 structure was slightly modified by either 0.005% sodium dodecyl sulfate or by the presence of octylglucoside (0.7%). Using this assay system, it was evident that proteolytic CYP2E1 activity was present both in the lumenal content of intact microsomal vesicles, as well as in the membranous "ghosts." Fractionation of the membranous activity on DEAE Sepharose revealed only one peak of activity, indicating the presence of only one or a restricted number of proteases active in CYP2E1 degradation. Also in another purification schedule based on initial hydroxyl apatite chromatography only one major peak was found, whereas a subsequent chromatography on DEAE Sepharose revealed three peaks of proteolytic activity, two which were active in CYP2E1 degradation. The proteases were finally purified by benzamidine Sepharose chromatography and sodium dodecyl sulfate–polyacrimde gel electrophoresis gels revealed one major band with M_r = 32000, which incorporated ^{3}H-DFP. The two proteases (microsomal cytochrome P450 protease 1 and 2, MCPP) had pH optima of 8 and were inhibited by agents specific for serine proteases. The proteolytic pattern of CYP2E1 obtained in the purified two-component system composed of CYP2E1 and protease was identical to the proteolytic system obtained in intact ghost membranes, yielding two major primary proteolytic fragments. Sequencing of these fragments revealed some specificity of the protease for lysine, with a recognition sequence of Lys-X- Lys (A. Zhukov et al., unpublished experiments). Experiments can now be designed to investigate whether the primary cleavage sites are affected by substrate binding. Perhaps the critical lysine residues are hidden in the three-dimensional structure upon substrate binding or become more exposed by P450 phosphorylation.

MCPP1 has now been sequenced in our laboratory in collaboration with Dr. Ulf Hellman, Ludwig Institute, Uppsala University, and been found to have a single amino acid sequence. Further cloning of this proteinase and isoforms thereof, as well as similar enzymes, will tell us about their roles in intracellular protein quality control, their tissue distribution, subcellular distribution, and their possible role in presentation of neoanti-genes for production of autoimmune antibodies, and for degradation of accumulated denatured proteins in various tissues, including the brain.

11.7 Alternative Pathways for Rapid P450 Degradation

It has been suggested that ubiquitin-dependent pathways are involved in the degradation of P450 enzymes (Tierney et al. 1992; Correia et al.

1992). Suicide substrates such as carbon tetrachloride for CYP2E1, and decarboethoxy-dimethyl-ethyl-dihydropyridine (DDEP) for CYP3A1 were used to trigger degradation. In the case of CYP2E1 disappearance of the protein was associated with increased formation of microsomal ubiquitin conjugates while fewer conjugates were formed after administration of 4-methyl pyrazol (Tierney et al. 1992). DDEP treatment caused rapid degradation of both CYP3A1 and CYP3A2, and the DDEP-related ubiquitin conjugates recognized by CYP3A1 antisera. In both these cases it appears that heme alkylation of the P450 is sufficient for ubiquitination. It remains to be established whether ubiquitin is conjugated with the intact P450 or to enzyme fragments previously generated by ER degradation, and whether the ubiquitination pathway, including heme alkylation as previously suggested (Correia et al. 1992), occurs under physiological conditions. A possibility is that phosphorylation of P450 triggers aggregation of the enzyme, and that such a force could provide substrates for soluble and ER-bound (Sommer and Jentsch 1993) ubiquitin-dependent proteolytic systems, as well as for non-ubiquitin-dependent systems in the ER.

11.8 Conclusions

It is evident that substrate interactions with cytochrome P450 directly determine the half-life of the protein and the intracellular routes for sorting and degradation of at least some forms of cytochrome P450. One crucial point of substrate action appears to be a protection of the enzymes from cAMP-dependent phosphorylation leading to heme loss and rapid degradation. Figure 1 summarizes our current model for this substrate-dependent regulatory process. The future challenges of this area of research definitely include the characterization of the membranous proteolytic systems able to distinguish between properly folded and misfolded variants of P450 and other proteins.

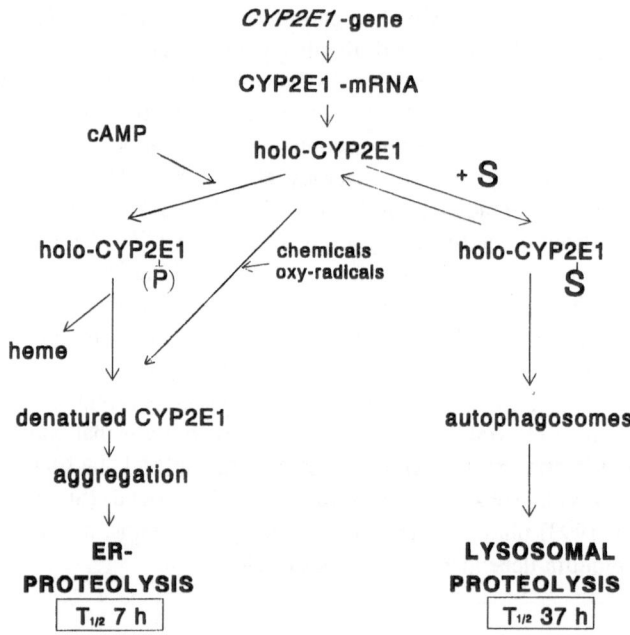

Fig. 1. Hypothetical scheme for posttranslational regulation of CYP2E1

References

Amara J F, Lederkremer G, Lodish HF (1989) Intracellular degradation of un-
    assembled asialoglycoprotein receptor subunits: a pre-Golgi, nonlysosomal
    endoproteolytic cleavage. J Cell Biol 109:3315–3324
Badger TM, Huang J, Ronis M, Lumpkin CK (1993) Induction of cytochrome
    P450 2E1 during chronic ethanol exposure occurs via transcription of the
    CYP2E1 gene when blood alcohol concentrations are high. Biochem
    Biophys Res Commun 190:780–785
Baker RK, Lively MO (1987) Purification and characterization of the oviduct
    microsomal signal peptidase. Biochemistry 26:8561–8567
Blanck J, Ristau O, Zhukov AA, Archakov AI, Rein H, Ruckpaul K (1991)
    Cytochrome P-450 spin state and leakiness of the monooxygenase path-
    way. Xenobiotica 21:121–135
Bonifacino JS, Suzuki CK, Klausner RD (1990) A peptide sequence confers
    retention and rapid degradation in the endoplasmic reticulum. Sience
    247:79–82

Bonifacino JS, Lippincott-Schwartz J (1991a) Degradation of proteins within the endoplasmic reticulum. Curr Opin Cell Biol 3:592–600

Bonifacino JS, Cosson P, Shah N, Klausner RD (1991b) Role of potentially charged transmembrane residues in targeting proteins for retention and degradation within the endoplasmic reticulum. EMBO J 10:2783–2793

Brennan SO, Peach RJ (1988) Calcium-dependent KEX2-like protease found in hepatic secretory vesicles converts proalbumin to albumin. FEBS Lett 229:167–170

Cheng SH, Gregory RJ, Marshall J, Paul S, Souza DW, White GA, O'Riordan CR, Smith AE (1990) Defective intracellular transport and processing of CFTR is the molecular basis of most cystic fibrosis. Cell 63:827–834

Chun KT, Bar NS, Simoni RD (1990) The regulated degradation of 3-hydroxy-3-methylglutaryl-CoA reductase requires a short-lived protein and occurs in the endoplasmic reticulum. J Biol Chem 265:22004–22010

Claudio T, Paulson HL, Green WN, Ross AF, Hartman DS, Hayden D (1989) Fibroblasts transfected with Torpedo acetylcholine receptor β- χ-, and δ-subunit cDNAs express functional receptors when infected with a retroviral recombinant. J Cell Biol 108:2277–2290

Correia MA, Davoll SH, Wrighton SA, Thomas PE (1992) Degradation of rat liver cytochromes P450 3A after their inactivation by 3,5-dicarbethoxy-2,6-dimethyl-4-ethyl-1,4-dihydropyridine: characterization of the proteolytic system. Arch Biochem Biophys 297:228–238

Cotner T (1992) Unassembled HLA-DR β monomers are degraded rapidly by a nonlysosomal mechanism. J Immunol 148:2163–68

Doyle C, Sambrook J, Gething M-J (1986) Analysis of progressive deletions of the transmembrane and cytoplasmic domains of influenza virus hemagglutinin. J Cell Biol 103:1193–1204

Eliasson E, Johansson I, Ingelman-Sundberg M (1988) Ligand-dependent maintenance of ethanol-inducible cytochrome P-450 in primary rat hepatocyte cell cultures. Biochem Biophys Res Commun 150:436–443

Eliasson E, Johansson I, Ingelman-Sundberg M (1990) Substrate, homone and cAMP- regulation of cytochrome P450 degradation. Proc Natl Acad Sci USA 87:3225–3229

Eliasson E, Mkrtchian S, Ingelman-Sundberg M (1992) Hormone- and substrate regulated intracellular degradation of cytochrome P450 (2E1) involving MgATP-activated rapid proteolysis in the endoplasmic reticulum membranes. J Biol Chem 267:15765–15769

Eliasson E, Mkrtchian S, Halpert J, Ingelman-Sundberg M (1994) Substrate-regulated cAMP-dependent phosphorylation, aggregation and degradation of glucocorticoid-inducible rat liver cytochrome P450 3A1. J Biol Chem (accepted for publication)

Esser V, Russell DW (1988) Transport-deficient mutations in the low density lipoprotein receptor. Alterations in the cysteine-rich and cysteine-poor regions of the protein block intracellular transport. J Biol Chem 263:13276-13281

Evans EA, Gilmore R, Blobel G (1986) Purification of microsomal signal peptidase as a complex. Proc Natl Acad Sci USA 83:581–585

Gardner AM, Aviel S, Argon Y (1993) Rapid degradation of an unassembled immunoglobulin light-chain is mediated by serine protease and occurs in a pregogli compartment. J Biol Chem 268:25940–25947

Goldstein JL, Brown MS (1990) Regulation of the mevalonate pathway. Nature 343:425–430

Gorsky LD, Koop DR, Coon MJ (1984) On the stoichiometry of the oxidase and monooxygenase reactions catalyzed by liver microsomal cytochrome P-450. J Biol Chem 259:6812–6817

Gotoh O (1992) Substrate recognition sites in cytochrome P450 family 2 (CYP2) proteins inferred from comparative analysis of amino acid and coding nucleotide sequences. J Biol Chem 267:83–90

Hanukoglu I, Rapoport R, Weiner L, Sklan D (1993a) Electron leakage from the mitochondrial NADPH-adrenodoxin reductase-adrenodoxin-P450scc (cholesterol side chain cleavage) system. Arch Biochem Biophys 305:489498

Hanukoglu I, Rapoport R, Vonrhein C, Raikhinstein M, Sklan D, Schulz G (1993b) Regulation of electron flow in the mitochondrial cytochrome P450 systems. Abstracts of the 8th International Conference of Cytochrome P450: biochemistry, biophysics, and molecular biology, Lisbon, 24–28 Oct 1993, pp 21–22

Hoe MH, Hunt RC (1992) Loss of one asparagine-linked oligosaccharade from human transferrin receptor results in specific cleavage and association with the endoplasmic reticulum. J Biol Chem 267:4916–4923

Hostetler KA, Wrighton SA, Malowa DT, Thomas PE, Levin W, Guzelian PS (1989) Coinduction of multiple hepatic cytochrome P450 proteins and their mRNAs in rats treated with imidazole antimycotic agents. Mol Pharmacol 35:279–285

Johansson I, Scholte B, Eliasson E, Ingelman-Sundberg M (1988) Ethanol-, fasting- and acetone-inducible cytochromes P-450 in rat liver. Pergamon, New York, pp 231-235 (Advances in biosciences, vol 71)

Johansson I, Lindros KO, Eriksson H, Ingelman-Sundberg M (1990) Transcriptional control of CYP2E1 in the perivenous liver region and during starvation. Biochem Biophys Res Commun 173:331–338

Johansson I, Eliasson E, Ingelman-Sundberg M (1991) Hormone-controlled phosphorylation and degradation of CYP2B1 and CYP2E1 in isolated rat hepatocytes. Biochem Biophys Res Commun 174:37–42

Karuzina II, Archakov AI (1994) The oxidative inactivation of cytochrome P450 in monooxygenase reactions. Free Radic Biol Med 16:73–97

Kishimoto TK, Hollander N, Roberts TM, Anderson DC, Springer TA (1987) Heterogenous mutations in the β subunit common to the LFA-1, Mac-1, and p150,95 glycoproteins cause leukocyte adhesion deficiency. Cell 50:193–202

Klausner RD, Sitia R (1990) Protein degradation in the endoplasmic reticulum. Cell 62:611–614

Lau MMH, Neufeld EF (1989) A frame-shift mutation in a patient with Tay-Sachs disease causes premature termination and defective intracellular transport of the α subunit of β-hexodaminidase. J Biol Chem 264:21376-21380

Le A, Graham KS, Siefers RN (1990) Intracellular degradation of the transport impaired human PeZ α-antitrypsin variant. Biochemical mapping of the degradated event among compartments of the secretory passway. J Biol Chem 265:14001–14007

Le A, Ferrell GA, Dishon DS, Le Q-QA, Sifers RN (1992) Soluble aggregates of the human PiZ α1-antitrypsin variant are degraded within the endoplasmic reticulum by a mechanism sensitive to inhibitors of protein. J Biol Chem 267:1072–1080

Lippincott-Schwartz J, Bonifacino JS, Yuan LC, Klausner RD (1988) Degradation from the endoplasmic reticulum: disposing of newly synthesized proteins. Cell 54:209–220

Masaki R, Yamamoto A, Tashiro Y (1987) Cytochrome P450 and NADPH-cytochrome P450 reductase are degraded in the autolysosomes in rat liver. J Cell Biol 104:1207–1215

Nelson DR, Strobel HW (1988) On the membrane topology of vertebrate cytochrome P-450 proteins. J Biol Chem 263:6038–6050

Park KS, Sohn DH, Veech RI, Song BJ (1993) Translational activation of ethanol-inducible cytochrome (CYP2E1) by isoniazid. Eur J Pharmacol 248:7–14

Porter TD, Coon MJ (1991) Cytochrome P450. Multiplicity of isoforms, substrates, and catalytic and regulatory mechanisms. J Biol Chem 266:13469-13472

Poulus TL, Finzel BS, Gunzalus IC, Wagner GC, Kraut J (1985) The 2.6-Å crystal structure of Pseudemonas putida cytochrome P-450. J Biol Chem 265:16122–16130

Pyerin W, Taniguchi H (1989) Phosphorylation of hepatic phenobarbital-inducible cytochrome P450. EMBO J 8:3003–3010

Ronis MJJ, Johansson I, Hultenby K, Lagercrantz J, Glaumann H, Ingelman-Sundberg M (1991) Acetone regulated synthesis and degradation of cytochromes P450IIE1 and P450IIB1 in rat liver. Eur J Biochem 198:383–389

Ronis MJJ, Lumpkin CK, Ingelman-Sundberg M, Badger TM (1991) Effects of short-term ethanol and nutrition on the hepatic microsomal monooxygenase system in a model utilizing total enteral nutrition in the rat. Alcohol Clin Exp Res 15:693–699

Ronis MJJ, Crough J, Mercado C, Irby D, Valentine C, Lumpkin CK, Ingelman-Sundberg M, Badger TM (1993) Cytochrome P450 CYP2E1 induction during alcohol exposure occurs by a two step mechanism associated with blood alcohol concentrations in rats, J Pharm Exp Ther 264:944–950

Rotundo RL (1988) Biogenesis of acetylcholinesterase molecular forms in muscle: evidence for a rapidly turning over, catalytically inactive precursor pool. J Biol Chem 263:19398–19406

Sato R, Imanaka T, Takatsuki A (1990) Degradation of newly-synthesized apolipoprotein B100 in a pre-Golgi compartment. J Biol Chem 265:11880-11884

Shamu CE, Cox JS, Walter P (1994) The unfolded-protein response pathway in yeast. Trends Cell Biol 4:56–60

Sitia R, Neuberger MS, Molstein C (1987) Regulation of membrane IgM expression in secretory B cells: translational and post-translational events. EMBO J 6:3969–3977

Sligar SG, Murray RI (1986) Cytochrome P450cam and other bacterial P450 enzymes. In: Ortiz de Montellano PR (ed) Cytochrome P-450: structure, mechanism, and biochemistry. Plenum, New York, pp 429–503

Sommer T, Jentsch S (1993) A protein translocation defect linked to ubiquitin conjugation at the endoplasmic reticulum. Nature 365:176–179

Song BJ, Matsunaga T, Hardwick JP, Park SS, Veech RL, Yang CS, Gelboin HV, Gonzalez FJ (1987) Stabilization of cytochrome P450j messenger ribonucleic acid in the diabetic rat. Mol Endocrinol 1:542–547

Song BJ, Veech RL, Park SS, Gelboin HV, Gonzalez FJ (1989) Induction of rat hepatic N-nitrosodimethylamine demethylase by acetone is due to protein stabilization. J Biol Chem 264:3568–3572

Stoller TJ, Shields D (1989) The propeptide of preprosomatostatin mediates intracellular transport and secretion of α-globin from mammalian cells. J Cell Biol 108:1647–55

Tamanoe Y, Takahashi T, Takahashi K (1993) Purification and characterization of two isoforms of serine proteinase from the microsomal membranes of rat liver. J Biochem 113:229–235

Tierney DJ, Haas AL, Koop DR (1992) Degradation of cytochrome P450 2E1: selective loss after labilization of the enzyme. Arch Biochem Biophys 293:9–16

Tsao YS, Yvessa NE, Adesnick M, Sabatini DD, Kreibich G (1992) Carbox-ytermally truncated forms of rebophorin 1 are degraded in pregolgi com-partments by a calcium-dependent process. J Cell Biol 116:57–67

Tuckey RC, Kamin H (1982) The oxyferro complex of adrenal cytochrome P-450 scc. Effect of cholesterol and intermediates on its stability and optical characteristics. J Biol Chem 257:9309–9314

Urade R, Nasu M, Moriyama T, Wada K, Kiro M (1992) Protein degradation by the phosphoinositide-specific phospholipase C-α family from rat liver endoplasmic reticulum. J Biol Chem 267:15152–15159

Wang Y, Olson MJ, Baker MT (1993) Interaction of fluoroethane chlorofluo-rocarbon (CFC) substitutes with microsomal cytochrome P450. Stimula-tion of P450 activity and chlorodifluoroethene metabolism. Biochem Phar-macol 46:87–94

Wassler M, Fries E (1993) Proteolytic cleavage of haptoglobin occurs in the subcompartment of endoplasmic reticulum: evidence from membrane fu-sion vitro. J Cell Biol 123:285–291

Watkins PB, Wrighton SA, Scheutz EG, Maurel P, Buzelian PS (1986) Mac-rolide antibiotics inhibit the degradation of the glucocorticoid-responsive cytochrome P450p in rat hepatocytes in vivo and in primary monolayer culture. J Biol Chem 261:6264–6271

Wikström L, Lodish HF (1992) Endoplasmic reticulum degradation of a sub-unit of the asialoglycoprotein receptor in vitro. J Biol Chem 267:5–8

Wileman T, Kane LP, Young J, Carson GR, Terhorst C (1991) Associations between subunit ectodomains promote T cell antigen receptor assembly and protect against degradation in the ER. J Cell Biol 122:67–78

Zhukov AA, Archakov AI (1982) Complete stoichiometry of free NADPH ox-idation in liver microsomes. Biochem Biophys Res Commun 109:813–818

Zhukov AA, Archakov AI (1985) Stoichiometry of the reactions of microso-mal oxidation. Distribution of redox eqiovalents between monooxygenase and oxidase reactions catalyzed by cytochrome P-450. Biochemistry (USSR) 50:1659–1672

Zhukov A, Werlinder V, Ingelman-Sundberg M (1993) Purification and char-acterization of two membrane bound serine proteinases from rat liver microsomes active in degradation of cytochrome P450. Biochem Biophys Res Commun 197:221–228

12 Strategies for the Use of Single Cytochrome P450 Enzymes in Drug Research and Future Prospects

J. Doehmer

12.1   The Need for Genetically Engineered in Vitro Systems......... 213
12.2   Important Criteria for a Useful Development and Application
        of Cells Genetically Engineered for Cytochromes P450 214
12.3   The V79 Chinese Hamster Cell Battery 215
12.4   Relevance of Genetically Engineered V79 Cells 217
12.5   Future Prospects 220
References 222

12.1 The Need for Genetically Engineered In Vitro Systems

Analytical tools have always played an important role for the understanding of complex biological systems such as drug metabolism. Technology is required to develop analytical tools. In this context, cell culture technology has made a very important contribution to the area of in vitro systems (Doehmer 1993). The combination of gene technology and cell culture technology has provided new opportunities for studying drug metabolism related problems because any gene from any species encoding an enzyme relevant for drug metabolism may be cloned and expressed in bacterial, yeast, or mammalian cell in a defined way. This approach in drug metabolism is of particular importance because some of the enzymes are difficult to purify and to prepare in sufficient quantities, because of low expression levels, organ-specific expression, or low abundance of native organ material. These restrictions apply especially for human enzymes. The availibility of gene libraries for cloning the wanted gene as cDNA, followed by heterologous expression of this cDNA bypasses these restrictions. This approach also substitutes drug metabolism studies in animals for drug safety studies in humans. Metabolism studies in animals are always biased by species differences. Thereby heterologous expression of relevant human genes facilitates drug metabolism studies even at the preclinical stage of drug development.

12.2 Important Criteria for a Useful Development and Application of Cells Genetically Engineered for Cytochromes P450

There are two critical steps in the gene technological approach to drug metabolism (a) the construction of the recombinant expression vector and (b) the selection of a cell for heterologous expression. These two steps determine the quality and usefulness of a genetically engineered in vitro system.

For constructing the recombinant expression vector several aspects must be considered. Most importantly, the expressed enzyme should be authentic in structure and function. The vector should therefore contain a full-length cDNA. The expression signal should be compatible with the host cells molecular biology for efficient transcription of the attached cDNA. Vector-controlled and -directed expression determine expression levels and thereby the detection limits of analytical devices.

The host cell for heterologous expression is best chosen when this cell has special characteristics and biological endpoints relevant for the metabolism-related problem to be studies. This decision requires detailed knowledge on the biology of the selected host cell. Badly characterized cells can spoil the advantages of heterologous expression. This is particularly the case when cells are selected on the basis of being close to the in vivo situation because cells inevitably change their differentiation status upon cultivation. It is not fully understood what the mechanisms and controls are that maintain differentiation. For this reason several attempts to freeze the in vivo differentiation status in vitro by using, for example, large T-antigen have frequently failed (Pfeifer et al.

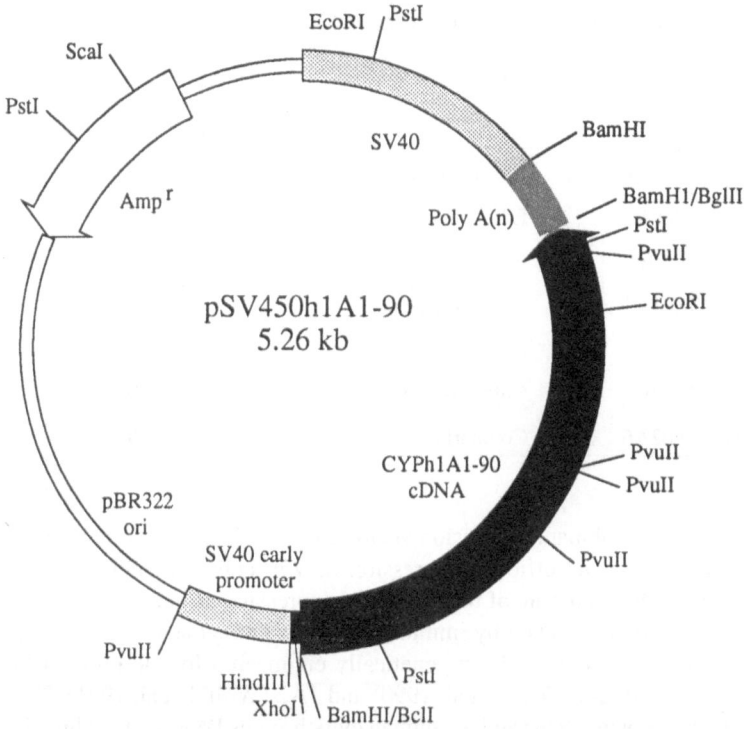

Fig. 1. Recombinant cytochrome P450 plasmid routinely used for heterologous expression in V79 cells (Schmalix et al. 1993)

1993). For clarity and to avoid confusions, unfounded arguments, and overstatements it is important to consider in vitro systems as analytical tools only. This goal is best served the better a system is defined, and the easier the cell is handled, regardless of whether the cell is of human or animal origin.

12.3 The V79 Chinese Hamster Cell Battery

The above criteria underline the strategy for genetically engineering V79 for stable expression of cytochromes P450.

Table 1. Activities in genetically engineered V79 Chinese hamster cells

Cell line	Substrate	Specific activity (pmol min^{-1}, mg^{-1} total protein)
V79MZr1A1	Benzo(*a*)pyrene	48
V79MZr1A2	Ethoxyresorufin	8
V79MZr2B1	Pentoxyresorufin	30
V79MZh1A1	Benzo(*a*)pyrene	48
V79MZh1A2	Ethoxyresorufin	5
V79MZh2E1	Chlorzoxazone	30
V79MZh2A6	Coumarin	80

The recombinant expression vector contains the SV40 early promotor for stable and efficient expression in V79 cells. The cDNAs contained in the vector are of full length for expressing an authentic cytochrome P450, as checked by immunological and enzyme assays (Fig. 1). So far V79 cells have been genetically engineered for the rat cytochromes P450 1A1 (Dogra et al. 1990) and 1A2 (Wölfel et al. 1991), 2B1 (Doehmer et al. 1988) and the human cytochromes P450 1A1 (Schmalix et al. 1993), 1A2 (Wölfel et al. 1992), 2E1 (in preparation), and 2A6 (in preparation). In addition, V79 cells have also been genetically engineered for the human cytochrome P450 reductase (in preparation; Table 1).

V79 Chinese hamster cells were initially selected as host cell for heterologous expression of cytochromes P450 because these cells are widely used in various toxicological testing, such as mutagenicity, micronucleus formation, sister chromatid exchange. V79 cells have some advantages, including no special nutrient requirement, rapid growth, high cloning efficiency, and stable karyotype. For studying cytochrome P450 mediated effects V79 cells had to be applied in conjunction with liver homogenate because V79 do not express cytochromes P450. This has been confirmed repeatedly in our studies. No endogenous cytochrome P450 mRNA, protein, or activity has ever been detected in V79 cells even under cytochrome P450 inducible conditions.

There is only one report describing a cytochrome P450 1A1 related activity at 0.03 pmol min^{-1} mg^{-1} total protein in V79 cells (Kiefer and Wiebel 1989), which is in most studies below detection limit. Therefore, V79 cells can be expected to be void of cytochromes P450. For this reason, genetically engineered V79 cells are defined for the cDNA encoded cytochrome P450.

12.4 Relevance of Genetically Engineered V79 Cells

Cytochrome P450 expressing V79 cells have been applied in a variety of metabolism-related toxicological studies. These include mutagenicity (Glatt et al. 1993), cytotoxicity (Schmalix et al. 1993), chromosomal aberration (Kulka et al. 1993; Jensen et al. 1993a), micronucleus formation studies (Ellard et al. 1991), tumor promotion (Vang et al. 1993), investigation a variety of chemicals such as polycyclic aromatic hydrocarbons and nitrosamines. Those studies have revealed interesting species differences on cytochrome P450 dependent metabolic activation of procarcinogens which are of relevance for risk assessment.

In the meantime cytochrome P450 expressing V79 cells have also been shown to be valuable tools for studying drug metabolism (Doehmer et al. 1990; Fuhr et al. 1992; Jensen et al. 1993b). For example, a comparative study revealed a cytochrome P450 1A2 species-specific demethylation pattern of caffeine. Using caffeine metabolites as metabolic markers, V79 cells expressing cytochromes P450 1A2 are being used in drug interaction studies. Drugs given simultaneously may influence their metabolism if one of these drugs functions as a cytochrome P450 inhibitor causing a lack of metabolic activation of another drug.

Phenacetin-O-deethylation was measured comparatively in V79 cells expressing rat cytochrome P450 1A2 and in primary rat hepatocytes. In both cells phenacetin was effectively changed to paracetamol. However, a striking difference was observed in enzyme kinetic studies. Due to the presence of many different cytochromes P450 there is no linear dose response curve in the Eadie Hofstee plot due to low-affinity binding sites for phenacetin with no catalytic activity but in V79 cells showed a linear relationship allowing the evaluation of cytochrome P450 1A2 related kinetic data (Fig. 2). In this particular case it was of advantage to apply genetically engineered V79 cells rather than primary

Table 2. Mechanism-based in vitro systems: advantages and disadvantages

Primary hepatocytes	Close to in vivo situation
	Inducible expression of CYPs
	Rapid loss of CYPs
	Loss of CYPs at different rates
	CYPs expressing cells do not grow
	Limited availability
Established cell lines	Unlimited availability
	Useful biological endpoints
	Some CYP activity
	Stable expression
	Inducible expression of CYPs
	Unknown CYP activity
	Lack of most CYPs
Genetically engineered cell lines	Unlimited availability
	Useful biological endpoints
	Highly defined for CYP activity
	Stable expression
	Rapid growth
Subcellular fractions	Unlimited/limited availability
	No biological endpoints
Purified enzymes	Unlimited/limited availability
	Highly defined for CYP activity
	No biological endpoints

hepatocytes. However, there is no such thing as the ultimate in vitro system. Advantages and disadvantages of the various in vitro possibilities must be weighted for their usefulness to provide answers and solutions to a specific problem (Table 2). Sometimes it sense to use a combination of different in vitro system. Although the use of primary hepatocytes for detailed kinetic studies is not advisable, these cells may function as a first screen for obtaining an indication of metabolically important cytochromes P450. Once the relevant cytochrome P450 isoform is identified, for example, by antibody inhibition studies, genetically engineered cells defined for this cytochrome P450 isoform may be applied for further and more detailed studies.

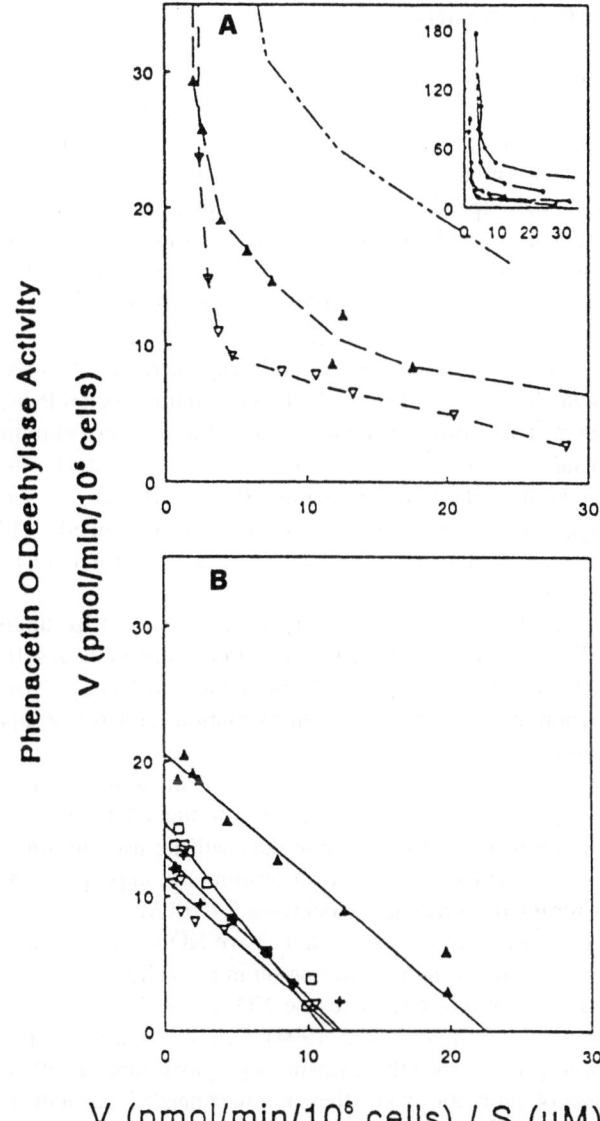

Fig. 2. Kinetics of phenacetin-O-deethylation in freshly prepared hepatocytes (A) and in V79r1A2 cells expressing rat CYP1A2 (B) (Jensen et al. 1993b)

12.5 Future Prospects

The SV40 early promotor directed expression of cytochromes P450 in V79 cells has been repeatedly shown to yield effective expression levels. The value of V79 cells as host cells for heterologous expression has been demonstrated in the various applications. Therefore it is worth continuing and expanding the V79 cell battery for drug-metabolizing enzymes. This should include not only cytochromes P450 but also coexpression with conjugating enzymes, as has been demonstrated for coexpression cytochrome P450 1A2 and acetyltransferase in mutagenicity studies of aromatic amines (Fig. 3).

Recently V79 cells were successfully engineered for expressing the human cytochrome P450 11B1, which plays an important role in steroid metabolism. This isoform is a mitochondrial located cytochrome P450 and depends on reductase system different from the cytochrome P450 reductase in the endoplasmic reticulum. Cells were active as shown by 11-deoxycortisol hydroxylation (Denner et al., submitted). This will expand the usefulness of V79 cells as host cells for mitochondrial cytochromes P450.

New developments are currently being produced by linking the genetically engineered V79 cells (a) to molecular modeling for drug design, (b) important clinical problems related to NO synthase, toxic shock syndrome in sepsis and transplantation, and (c) occupational safety studies.

The inhibition potency of various drugs is measured using caffeine demethylation as metabolic marker in cytochrome P450 1A2 expressing V79 cells (Fuhr et al. 1992). These data will be used in drug design studies by molecular modeling to determine the importance of subtle changes for interactions with cytochrome P450 1A2.

Overproduction of NO due to a massive NO synthase induction by cytokines may impair biotransformation in critically ill patients suffering from massive infections because NO has an immediate effect on cytochromes P450 (Stadler et al. 1994). Drugs inhibiting the liver NO-synthase may be a protective measure for cytochromes P450 allowing high doses of antibiotics when they are most needed. As in the case for cytochromes P450 there is no NO synthase activity in V79 cells. V79 cells may be genetically engineered for human liver and other NO synthase isoforms for the study of NO synthase inhibitors.

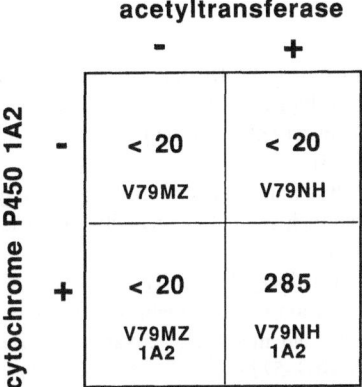

Fig. 3. Mutagenicity of 2-aminoanthracene as checked by HPRT assay in cells with no cytochrome P450 1A2 (V79MZ), with cytochrome P450 1A2 (V79MZr1A2), with acetyltransferase (V79NH), and with cytochrome P4501A2 together with acetyltransferase (V79NHr1A2)

For risk assessment studies regarding occupational safety and health it is important to know the cytochrome P450 dependent metabolite profile in general and individually. Individuals usually are checked for metabolites by noninvasive methods, for example, by breathing tests or checking urine samples. These metabolites may not be necessarily exactly the same as those which are initially formed in an organ. Comparing metabolite profiles obtained from genetically engineered V79 cells and those profiles as observed in urine samples may help to identify those metabolites which remain stable during excretion and may therefore be of increased diagnostic value.

These examples show a multidirectional application of V79 cells genetically engineered for cytochromes P450 and other enzymes. It is a matter of time until the V79 cell battery is complete, at least for the most important human cytochromes P450 related to drugs, which have been referred to as the big 11.

Acknowledgment. I gratefully acknowledge the generous support by the Bundesgesundheitsamt, Berlin, ZEBET, for the development of genetically engineered V79 cells.

References

Denner K, Vogel R, Schmalix W, Doehmer J, Bernhardt R (1994) Cloning and stable expression of human mitochondrial cytochrome P450 11B1 cDNA in V79 Chinese hamster cells and their application for testing of potential inhibitors (submitted)

Doehmer J (1993) V79 Chinese hamster cells genetically engineered for cytochrome P450 and their use in mutagenicity and metabolism studies. Toxicology 82:105–118

Doehmer J, Dogra S, Friedberg T, Monier S, Adesnik M, Glatt HR, Oesch F (1988) Stable expression of rat cytochrome P-450IIB1 cDNA in Chinese hamster cells (V79) and metabolic activation of aflatoxin B1. Proc Natl Acad Sci USA 85:5769–5773

Doehmer J, Glatt HR, Seidel A, Oesch F (1990) Genetically engineered V79 Chinese hamster cells metabolically activate the cytostatic drugs cyclophosphamide and ifosfamide. Environ Health Perspect 88:63–65

Dogra S, Doehmer J, Glatt HR, Mölders H, Siegert P, Friedberg T, Seidel A, Oesch F (1990) Stable expression of rat cytochrome P-450IA1 cDNA in V79 Chinese hamster cells and their use in mutagenicity testing. Molecular Pharmacology 37:608–613

Ellard S, Mohammed Y, Dogra S, Wölfel C, Doehmer J, Parry JM (1991) The use of genetically engineered V79 Chinese hamster cultures expressing rat liver CYP 1A1, 1A2, 2B1 cDNAs in micronucleus assays. Mutagenesis 6:461–470

Fuhr U, Doehmer J, Battula N, Wölfel C, Kudla C, Keita Y, Staib AH (1992) Biotransformation of caffeine and theophylline in mammalian cell lines genetically engineered for expression of single cytochrome P450 isoforms. Biochem Pharmacol 43:225–235

Glatt HR, Pauly K, Wölfel C, Dogra S, Seidel A, Lee H, Harvey RG, Oesch F, Doehmer J (1993) Stable Expression of heterologous cytochromes P450 in V79 cells: mutagenicity studies with polycyclic aromatic hydrocarbons. Polycycl Aromat Comp 3:1167–1174

Jensen KG, Önfelt A, Poulsen HE, Doehmer J, Loft S (1993a) Effects of benzo[a]pyrene and (±)-trans-7,8-dihydroxy-7,8-dihydrobenzo[a]pyrene on mitosis in Chinese hamster V79 cells with stable expression of rat cytochrome P450 1A1 or 1A2. Carcinogenesis 14:2115–2118

Jensen KG, Loft S, Doehmer J, Poulsen HE (1993b) Metabolism of phenacetin in V79 Chinese hamster cell cultures expressing rat liver cytochrome P450 1A2 compared to isolated rat hepatocytes. Biochem Pharmacol 45:1171-1173

Kulka U, Doehmer J, Glatt HR, Bauchinger M (1993) Cytogenetic effects of promutagens in genetically engineered V79 Chinese hamster cells expressing cytochromes P450. Eur J Pharmacol Environ Toxicol Pharmacol 228:299–304

Kiefer F, Wiebel FJ (1989) V79 Chinese hamster cells express cytochrome P-450 activity after simultaneous exposure to polycyclic aromatic hydrocarbons and aminophylline. Toxicol Lett 48:265–272

Pfeifer AMA, Cole KE, Smoot DT, Weston A, Groopman JD, Shields PG, Vignaud JM, Juillerat M, Lipsky MM, Trump BF, Lechner JF, Harris CC (1993) Simian virus 40 large tumor antigen-immortalized normal human liver epithelial cells express hepatocyte characteristics and metabolize chemical carcinogens. Proc Natl Acad Sci USA 90:5123–512

Schmalix WA, Mäser H, Kiefer F, Reen R, Wiebel FJ, Gonzalez F, Seidel A, Glatt HR, Greim H, Doehmer J (1993) Stable expression of human cytochrome P450 1A1 cDNA in V79 Chinese hamster cells and metabolic activation of benzo[a]pyrene. Eur J Pharmacol Environ Toxicol Pharmacol 248:251–261

Stadler J, Trockfeld J, Schmalix WA, Brill T, Siewert JR, Greim H, Doehmer J (1994) Inhibition of cytochromes P450 by nitric oxide. Proc Natl Acad Sci USA (in press)

Vang O, Wallin H, Doehmer J, Autrup H (1993) Cytochrome P450 mediated metabolism of tumor promoters modifies the inhibition of intercellular communication: a modified assay for tumor promotion. Carcinogenesis 14:2365-2371

Wölfel C, Platt KL, Dogra S, Glatt HR, Wächter F, Doehmer J (1991) Stable expression of rat cytochrome P450IA2 cDNA in V79 Chinese hamster cells and hydroxylation of 17β-estradiol and 2-aminofluorene. Mol Carcinog 4:489–498

Wölfel C, Heinrich-Hirsch B, Seidel A, Frank H, Ramp U, Wächter F, Wiebel F, Gonzalez F, Doehmer J (1992) Genetically engineered V79 Chinese hamster cells for stable expression of human cytochrome P450IA2 cDNA. Eur J Pharmacol Environ Toxicol Pharmacol 228:95–102

Index

abecarnil 11
adenyl cyclase 199
adrenaline 199
adrenodoxin 86
adrenodoxin reductase 86
aflotoxin B_1 170
Ah-receptor 28, 36
allosteric site 173
anticancer drug 59
autophagosomal-lysosomal
    pathway 201

bacteria 85
bacterial expression system 93
bacterial membrane 87
barbiturate 163
7,8-benzoflavone 173
benzo(a)pyrene 173

cDNA expression 112, 115, 116,
    123, 125, 128
chemical inhibitor 52
cholesterol side-chain cleavage
    cytochrome P4 82
cicaprost 5
circulating antibodies 173
cirrhosis 164
clinical consequence 36
cortisol 168

cyclophosphamide 57–59, 61–64
    66, 67, 70–74, 76–78
cyclosporin 164
cyclosporin A 175
CYP2B1 59, 61, 63, 66, 67, 70,
    72–74, 76, 77, 78, 198
CYP2E1 196
CYP3A1 199
cytochrom b_5 97, 100, 106, 107, 164
cytochrome P450 43–45, 49, 58,
    59, 61, 64, 66, 70, 72, 77, 78, 135,
    136, 145, 150, 152, 188, 216–218,
    220, 221
cytochrome P450 3A 161

dapsone 171
designer P450s 88
desogestrel 11
dexamethasone 163
1,4-dihydropyridines 165
disease 44, 45
drug and pollutant metabolism 98,
    108
drug metabolism 105, 136, 138,
    144, 152, 153

E. coli 82, 164
enzyme stabilization 196
epoxide hydrolase 98, 107, 108
eptalprost 5

ER-degradation   202
ergolines   7
erythromycin   171
estradiol   9
17α-Estradiol   168
ethinylestradiol   11
17α-Ethynylestradiol   168
evolution   24–26
extrachromosomal vector   112,
    115, 121

felodipine   173
ferrodoxin   86
– reductase   86
flavodoxin   89
– reductase   89
fusion protein   165

gedocarnil   11
gene therapy   70, 71, 77, 78
genetic polymorphism   44, 45, 51,
    53, 54
gestodene   170
glucagon   199
glutathione   165
Golgi apparatus   201
grapefruit juice   172

hepatocytes   4, 7, 14
heterologous expression   214, 216,
    220
history   21
human cell   112, 114
human fetal liver   163
human liver   188–191
human P450   100, 106
human p450 reductase   101, 105
human P450s   27, 108
6β-hydroxylation   171
17α-hydroxylase cytochrome
    P450   83

ifosphamide   57–59, 61–64, 66, 67,
    71, 72
iloprost   5
immunotoxicology   191
in vitro cDNA expression   111
in vitro prediction   52
induction   44
inhibition   45, 52
interindividual variation   49, 53, 54

ketoconazole   170

lidocaine   171
lisuride   7
liver bank   50, 51
liver perfusion   4, 7
liver slice   4, 14
lysosomes   201

mammalian cell   154
metabolic interaction   3
metabolism   1, 213, 214, 217
methylation   12
microsomal P450s   83
microsomes   4, 14
midazolam   171
mitochondrial P450s   82
mRNA stabilization   196

NADPH cytochrome P450
    reductase   86
α-naphtholavone   173
nifedipine   164
nomenclature   23, 24

oxidase activities   197

P450 1A1   9
P450 1A2   9
P450 2B1   9
P450 reductase   99, 103–108

P450s 3A3, 3A4, 3A5, 3A7   163
peroxisome proliferator activated receptor   29
phase I reaction   1, 14
phase II reaction   2, 14
phenobarbital   30, 33
polymorphisms   26
posttranscriptional regulation   34
posttranslational stabilization   197
progesterone   168
prostacyclin   5

receptor-mediated regulation   28
recombinant   137, 150, 152, 153
regulation   24, 27, 30, 31, 33, 36
rifampicin   163

screening tool   9
secretorial pathways   201
site-directed mutagenesis   90

species differences   24, 25, 27
steroid hormones   25, 30
structure   21, 23, 25

terguride   7
testosterone   164
tissue-specific regulation   31
translation efficiency   196
troleandomycin   170
ubiquitin conjugates   205
urinary alcohol   197

V79 cells   9

xenobiotic metabolism   188, 191,
   192

yeast expression   98, 101
yeast Saccharomyces
   cerevisiae   164